Effective Implementation of an ISO 50001 Energy Management System (EnMS)

Also available from ASQ Quality Press:

Sustainable Business and Industry: Designing and Operating for Social and Environmental Responsibility
Joseph Jacobsen

An Introduction to Green Process Management
Sam Windsor

ANSI/ISO/ASQ E14001-2004: Environmental management systems—Requirements with guidance for use
ANSI/ISO/ASQ

The ASQ Supply Chain Management Primer
ASQ's Customer-Supplier Division and J.P. Russell, editor

The ASQ Quality Improvement Pocket Guide: Basic History, Concepts, Tools, and Relationships
Grace L. Duffy, editor

The ASQ Pocket Guide to Root Cause Analysis
Bjørn Andersen and Tom Natland Fagerhaug

The Quality Toolbox, Second Edition
Nancy R. Tague

The Certified Six Sigma Green Belt Handbook
Roderick A. Munro, Matthew J. Maio, Mohamed B. Nawaz, Govindarajan Ramu, and Daniel J. Zrymiak

The Certified Manager of Quality/Organizational Excellence Handbook, Fourth Edition
Russell T. Westcott, editor

The Certified Six Sigma Black Belt Handbook, Second Edition
T.M. Kubiak and Donald W. Benbow

The ASQ Auditing Handbook, Fourth Edition
J.P. Russell, editor

Root Cause Analysis: Simplified Tools and Techniques, Second Edition
Bjørn Andersen and Tom Fagerhaug

To request a complimentary catalog of ASQ Quality Press publications, call 800-248-1946, or visit our website at http://www.asq.org/quality-press.

Effective Implementation of an ISO 50001 Energy Management System (EnMS)

Marvin T. Howell

ASQ Quality Press
Milwaukee, Wisconsin

American Society for Quality, Quality Press, Milwaukee 53203
© 2014 by ASQ
All rights reserved.
Printed in the United States of America
19 18 17 16 15 14   5 4 3 2 1

Library of Congress Cataloging-in-Publication Data
Howell, Marvin T., 1936–
  Effective implementation of an ISO 50001 energy management system (EnMS) / Marvin T. Howell.
    pages cm
  Includes bibliographical references and index.
  ISBN 978-0-87389-872-0 (alk. paper)
  1. Industries—Energy conservation. 2. Industries—Energy consumption. 3. Management—Environmental aspects. I. Title.
  TJ163.3.H684 2014
  658.2'60218—dc23

                                                                    2013047399

No part of this book may be reproduced in any form or by any means, electronic, mechanical, photocopying, recording, or otherwise, without the prior written permission of the publisher.

Acquisitions Editor: Matt Meinholz
Managing Editor: Paul Daniel O'Mara
Production Administrator: Randall Benson

ASQ Mission: The American Society for Quality advances individual, organizational, and community excellence worldwide through learning, quality improvement, and knowledge exchange.

Attention Bookstores, Wholesalers, Schools, and Corporations: ASQ Quality Press books, video, audio, and software are available at quantity discounts with bulk purchases for business, educational, or instructional use. For information, please contact ASQ Quality Press at 800-248-1946, or write to ASQ Quality Press, P.O. Box 3005, Milwaukee, WI 53201-3005.

To place orders or to request a free copy of the ASQ Quality Press Publications Catalog, visit our website at http://www.asq.org/quality-press.

∞ Printed on acid-free paper

Quality Press
600 N. Plankinton Ave.
Milwaukee, WI 53203-2914
E-mail: authors@asq.org
The Global Voice of Quality℠

INFOTECH STANDARDS INDIA PVT. LTD.
4760-61, 2nd Floor, "SAI SAROVAR"
23, Ansari Road, Daryaganj
New Delhi - 110002, INDIA
www.standardsmedia.com
E-mail : info@standardsmedia.com
This edition is authorized for sale in India.
First Impression, 2015   ISBN 978-81-748-9030-6

This book is dedicated to the environmental and energy staff of the Drug Enforcement Administration (DEA) and the staff and special agents in the field who perform diligently and with exceptional dedication every day in achieving their mission. This book is also dedicated to the United States Air Force, where I had the privilege of serving 20 years in the civil engineering field and retiring as a lieutenant colonel. The Air Force is dedicated to reducing energy use and is making excellent strides in doing so, including creating an energy savings culture where everyone is committed to and involved in reducing energy consumption and costs. And finally, this book is dedicated to my wife, Jackie Howell, and my oldest grandson, Christopher Cline, who helped me with the figures and tables.

Table of Contents

List of Figures and Tables ix
Preface xiii

**Chapter 1   Introduction to ISO 50001 Environmental Management
    System (EnMS)** **1**
  Energy Cost 1
  EnMS Standard 1
  Integration of ISO Standards 2
  A Hypothetical Company, QVS Corporation, Is Used to Demonstrate
      Implementation 2
  EnMS Phases and Elements 3
  EnMS Stages 4

Chapter 2   Requirements **7**
  General Requirements 7
  Management Responsibilities 8
  Management Representative 10
  A Best Practice: Using OAR and/or PAL for Energy Team Meetings 13
  Summary of Energy Team Meetings and Meeting Metrics 16

Chapter 3   Energy Policy **21**
  Corporate Vision for Energy 21

Chapter 4   Energy Planning **25**
  An Overview 25
  General Requirements 26
  Legal and Other Requirements 26
  Energy Review 28
  Identifying Energy Efficiencies 37
  Lean Energy Analysis 40
  Identifying Energy Variables 43
  Energy Baseline 46
  EnPIs 47
  Energy Objectives, Energy Targets, and Energy Management
      Action Plans 51
  Energy Action Plans and Projects 60

Chapter 5 Implementation and Operations	**65**
An Overview	65
Competence, Training, and Awareness	66
Communications	70
Documentation	73
Control of Records	75
Operational Controls	76
Design	79
Procurement of Energy Services, Products, and Equipment	84
Chapter 6 Checking	**85**
Monitoring, Measurement, and Analysis	85
Evaluation of Compliance with Legal Requirements	96
Internal Audit of EnMS	99
Nonconformities, Corrective Actions, and Preventive Actions	109
Control of Records	114
Chapter 7 Management Review	**115**
General Requirements	115
Inputs to Management Review	116
Outputs from Management Review	118
Chapter 8 Integration of ISO Standards	**125**
Integration	125
QVS Corporation Integration of ISOs	125
Chapter 9 Pitfalls and Countermeasures	**147**
Murphy's Law	147
Possible Pitfalls and Countermeasures	147
Pitfall Observers	152
Chapter 10 Implementing ISO 50001 EnMS in Four Months or Less	**153**
Introduction	153
Proposed Schedule	153
Appendix A QVS Corporation Management Review	**157**
Appendix B List of Acronyms	**161**
Endnotes	*163*
Glossary	*165*
Bibliography	*171*
Index	*173*

List of Figures and Tables

Figure 1.1	How does an EnMS work?	3
Figure 1.2	The EnMS elements	4
Figure 1.3	The EnMS stages	5
Figure 2.1	Sample meeting agenda	14
Figure 2.2	Sample meeting minutes	15
Table 2.1	Summary of energy team meetings and metrics	17
Figure 3.1	Sample energy policy	23
Figure 3.2	QVS Corporation energy team's draft energy policy	23
Table 3.1	Evaluation/check of the energy policy	24
Figure 4.1	QVS Corporation sample self-inspection checklist	27
Figure 4.2	SEU selection process	30
Figure 4.3	Example of stratifying: HVAC	30
Figure 4.4	Example review	31
Table 4.1	QVS Corporation 2010 electricity profile	31
Figure 4.5	QVS Corporation 2010 utility costs	32
Figure 4.6	QVS Corporation HQ facility 2010 electricity profile	32
Figure 4.7	QVS Corporation Plant A 2010 electricity profile	33
Figure 4.8	QVS Corporation Plant B 2010 electricity profile	33
Figure 4.9	QVS Corporation Plant C 2010 electricity profile	34
Figure 4.10	QVS Corporation Plant D 2010 electricity profile	34
Table 4.2	Stratifying SEUs	35
Table 4.3	QVS Corporation's SEUs	36
Table 4.4	QVS Corporation's SEU energy variables	44
Table 4.5	QVS Corporation 2010 electricity baseline for all facilities	47
Figure 4.11	Sample data collection plan	48
Figure 4.12	QVS Corporation first EnPI	49
Figure 4.13	QVS Corporation second EnPI	49
Figure 4.14	QVS Corporation third EnPI	50
Figure 4.15	QVS Corporation fourth EnPI	51
Table 4.6	Sample form for evaluation and selection of possible objectives	54
Figure 4.16	Objective, target, and action plan template	55

Table 4.7	Implementation requirements being considered as an objective	57
Table 4.8	QVS Corporation implementation requirements being considered as an objective	57
Figure 4.17	QVS Corporation O&T action plan	59
Figure 4.18	QVS Corporation sample payback estimate	61
Table 4.9	QVS Corporation funded projects	61
Table 5.1	Sample organization's training plan/matrix	68
Table 5.2	QVS Corporation training plan/matrix	69
Figure 5.1	QVS Corporation's communication plan	71
Figure 5.2	QVS Corporation's documentation plan	75
Table 5.3	QVS Corporation's SEU operational controls—electricity	77
Table 5.4	QVS Corporation's SEU operational controls—natural gas	78
Figure 5.3	QVS Corporation's energy contingency plan	79
Figure 5.4	Sample ECM	80
Table 5.5	QVS Corporation's electricity audit (five facilities)	82
Figure 5.5	QVS Corporation ECM	83
Figure 5.6	Process variables	83
Figure 5.7	QVS Corporation's procurement plan	84
Figure 6.1	Power factor reading at QVS Corporation Plant B	87
Figure 6.2	Power factor cumulative savings at QVS Corporation Plant B	87
Table 6.1	EnPIs	88
Figure 6.3	QVS Corporation's monitoring and measurement guide	90
Figure 6.4	CSF measurement and assessment tool	92
Table 6.2	First use of CSF assessment tool (March 2011)	94
Table 6.3	CSF assessments	94
Figure 6.5	QVS Corporation 2011 CSF scores	95
Figure 6.6	QVS Corporation 2012 CSF scores	96
Table 6.4	Legal and other requirements compliance evaluation, 2012 (short form)	97
Figure 6.7	Legal and other requirements compliance evaluation, 2012 (long form)	98
Figure 6.8	Sample internal audit/self-inspection checklist	100
Figure 6.9	QVS Corporation EnMS internal audit/self-inspection checklist	105
Figure 6.10	Typical CAR	110
Figure 6.11	QVS Corporation energy team profile worksheet	112
Figure 6.12	QVS Corporation completed CAR	112
Figure 6.13	QVS Corporation's nonconformities and corrective and preventive action process	113
Figure 7.1	QVS Corporation management review agenda (January 15, 2013)	117
Figure 7.2	QVS Corporation management review minutes	120
Table 8.1	Evaluation of elements for possible integration	127
Figure 8.1	QVS Corporation environmental and energy stewardship policy	129

Figure 8.2	QVS Corporation's roles and responsibilities during team meetings	130
Table 8.2	Integrated documentation	132
Figure 8.3	QVS Corporation's integrated communication plan	134
Figure 8.4	QVS Corporation O&T action plan	137
Figure 8.5	QVS Corporation integrated CAR	138
Table 8.3	QVS Corporation's environmental and energy measurement and monitoring	140
Figure 8.6	QVS Corporation's integrated management review (EMS and EnMS) agenda	141
Figure 8.7	QVS Corporation's energy contingency plan	142
Table 8.4	QVS Corporation's emergency preparedness plan	143
Table 8.5	IST meeting summary	144
Table 10.1	Proposed implementation schedule	154

Preface

Several people have asked me, "Why do we need ISO 50001 Energy Management System (EnMS) when we have already implemented ISO 14001 Environmental Management System (EMS)?" Energy is part of an EMS. Energy is an aspect that is nonrenewable and a must for every organization to have. In ISO 14001 EMS, it is easy to focus on hazardous materials and aspects that have considerable risk in the workplace. Energy use can be easily overlooked, and even when it is considered for an objective and target (O&T), important questions such as what are the significant energy users (SEUs), what can we do to reduce their impact, and what are the variables that affect energy use are not answered. ISO 50001 EnMS allows an organization to focus on reducing energy consumption through establishing a compelling energy policy, establishing legal and other requirements and ensuring that they are being met, and conducting a comprehensive energy review that identifies energy efficiencies, energy conservation efforts implemented, and O&Ts with energy action plans that, when achieved, move the organization toward meeting its energy policy. For manufacturing companies, energy costs impact both the cost to produce the product and the product price. For government organizations, energy reduction is mandated by executive orders such as Executive Order 13423, "Strengthening Federal Environmental, Energy, and Transportation Management," and Executive Order 13514, "Federal Leadership in Environmental, Energy, and Economic Performance." Everyone benefits from reducing energy consumption, from the environment to the economic health of companies. ISO 50001 EnMS can be implemented by itself or with other ISO standards such as 9001 and 14001, or with OHSMS 18000. The choice is yours—let's make this a better place to live and work and with less cost.

ACKNOWLEDGMENTS

I would like to thank Acquisitions Editor Matt Meinholz and Managing Editor Paul Daniel O'Mara of ASQ Quality Press, and the ASQ reviewers who made contributions and suggestions that significantly improved the organization and clarity of this book. ISO is to be thanked for developing the ISO 50001 Energy Management System standard, which should help organizations all over the world reduce energy consumption and costs. I also thank Jana Brooks, former chief, Environmental Section, DEA Headquarters, who allowed me to assist in both environmental and energy matters at DEA facilities.

Chapter 1
Introduction to ISO 50001 Environmental Management System (EnMS)

ENERGY COST

In the United States, around $500 billion a year is spent on energy. In the world, industry consumes 51% of all energy produced. Energy costs represent up to 30% of corporate operating expenses.[1] The US Green Building Council estimates that commercial office buildings use, on average, 20% more energy than needed.[2] This is an astounding dollar loss for the industry that is due primarily to the fact that management does not know where the waste is occurring and what to do to eliminate or reduce this loss. By implementing an Energy Management System (EnMS), to include monitoring energy use, analyzing the data, and implementing countermeasures, companies can save hundreds of thousands of dollars each year. Transforming a company's facilities into high-performance buildings positively impacts the environment, employee productivity and well-being, and the company's bottom line.[3]

E_NMS STANDARD

The ISO 50001 EnMS standard was published in June 2011. ISO 50001 defines an EnMS as "a set of interrelated or interacting elements to establish an energy policy and energy objectives, and processes and procedures to achieve those objectives." The new standard used ISO 9001 Quality Management System (QMS) and ISO 14001 Environmental Management System (EMS) as guides in its development. ISO 50001 EnMS provides a road map and path for continually improving energy performance. In the past, projects to reduce energy use were identified, funded, and implemented. They were not usually tied to the organization's vision or strategic goals or objectives and involved only a few departments (e.g., engineering, contracting, and facilities personnel). This is often called a technical approach or method. ISO 50001 EnMS involves not only the technical but also an administrative or management approach in that top management and all the organization's employees and contractors are tasked to reduce energy use.[4]

Using an ISO standard as a guide for implementing your program has many benefits. First, it will help you identify opportunities to reduce energy use. It will assist you in putting appropriate operational controls in place. It will force you to understand your current energy usage and its related costs, and to look for ways to reduce your energy costs and consumption. It will help you gain management support and commitment and will help you explain to all the staff their roles and responsibilities. It will help you to be in better compliance with legal and other requirements. Your present metrics will be increased to help measure your total energy performance. It will enable you to put into practice procedures and processes to improve your design and procurement efforts in relationship to energy management. Most importantly, it will help you improve your energy performance, decrease your energy costs, and continually improve your EnMS. You may decide later to become certified as to conformance with the standard, after you have weighed the advantages and disadvantages of doing so.

INTEGRATION OF ISO STANDARDS

Thousands of organizations have implemented ISO 9001 QMS and/or ISO 14001 EMS. Many of the elements required in these two standards are similar to the requirements of ISO 50001 EnMS. Many companies implement ISO 50001 EnMS by consolidating it with their existing ISO 9000 or 14000 standards (integrating ISO 50001 EnMS with ISO 14001 is demonstrated in Chapter 8). By doing so, all requirements are covered, and maintenance of the two standards can be done more efficiently than maintaining each standard separately.

A HYPOTHETICAL COMPANY, QVS CORPORATION, IS USED TO DEMONSTRATE IMPLEMENTATION

There are two approaches for demonstrating how to plan, develop, implement, and maintain an EnMS. The first approach is to use several different types of companies and show what each did in its application. The second approach—and the one chosen by the author—is to create a hypothetical company that possesses several different types of facilities, thus symbolizing implementation in several different organizations. This will require creating specific requirements and documentation such as meeting agendas, minutes, objectives and targets (O&Ts), a communications plan, an energy policy, and other requirements. These documents can serve as guides for any organization in planning, developing, and implementing the ISO 50001 EnMS standard.

QVS Corporation is a company located in Gun Barrel City, Texas. Its facilities include a headquarters building with a large data center, a distribution center, and three plants with different missions and products produced. The structure of this company allows the author to show how to implement the standard in multiple locations where different issues are identified and overcome to achieve an excellent and effective EnMS. Applying the standard to these facilities provides experiences both with facilities that have seasonal electricity and natural gas usage and a facility that has almost constant energy use (the data center).

The structure followed in this book is straightforward:

1. Present the ISO 50001 EnMS requirement. (This is what the standard requires.)

2. Provide an operational explanation. (This is what the standard means and how any organization can implement processes to meet the intent and letter of the requirements.) This section describes which processes are necessary to conform to the standard and provides general guidance for all managers.

3. Include the forms, templates, and records needed. (Blank documents or templates to indicate data required to (a) implement the processes, plans, and process descriptions and (b) demonstrate completion of the processes described and how to document your activities.)

4. Develop and show the QVS Corporation example of implementation. (Examples of specifics include completed forms and templates showing how one company implemented processes to meet the requirements and how it documented the implementation.)

EnMS PHASES AND ELEMENTS

An EnMS has five phases, as shown in Figure 1.1. The five phases are supported with 23 elements, as shown in Figure 1.2.

To fully implement an EnMS, all 23 elements need to be achieved and documented with the information required by the standard. This book provides the standard's explanation of each of the 23 elements, the author's interpretation of what is required, and the forms, templates, plans, processes, and other meaningful information needed to comply with or satisfy the standard requirements.

Management commitment and involvement is an important part of meeting the standard. Developing a plan, implementing it, and checking for progress, results, or barriers are also major requirements. To simplify, ISO 50001 EnMS requires the following:

1. Management responsibilities

2. An energy policy

3. An energy process plan to achieve the energy policy

4. An implementation plan and execution

5. An evaluation or "check" to ensure everything is going as planned

6. Management reviews (take action)

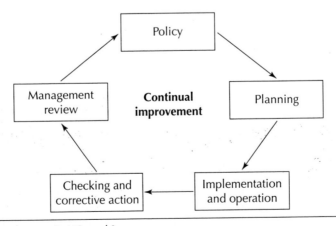

Figure 1.1 How does an EnMS work?

Source: "ISO 50001," *Wikipedia*, last modified October 29, 2013, http://en.wikipedia.org/wiki/ISO_50001/.

Chapter One

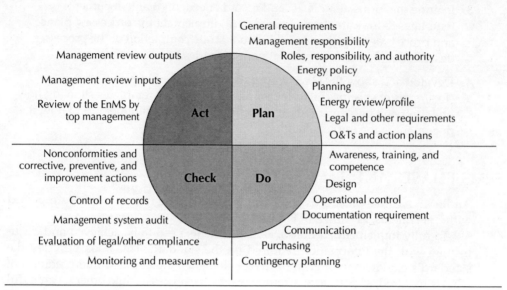

Figure 1.2 The EnMS elements.
Source: "ISO 50001," *Wikipedia*, last modified October 29, 2013, http://en.wikipedia.org/wiki/ISO_50001/.

These six areas comprise the plan-do-check-act cycle.[5] The book is organized to cover each of these.

ENMS STAGES

An EnMS has four stages, as shown in Figure 1.3. What is involved at each stage, and what is the composition of each stage? How is the sustaining stage different from the maintaining stage? These are important questions that are answered here.

The first stage is planning and development. In this stage, management commitment is received and an energy champion is appointed. An energy cross-functional team is assembled and put into place. The document system is established. The energy review is accomplished, achieving an energy profile, the scope, significant energy users (SEUs), SEUs' variables, and energy efficiencies. Procedures and required plans are developed. The time frame for this stage is normally six months, but it can be accelerated to be completed in one month.

The second stage is implementation. In this stage the O&Ts are implemented. Meetings are held by the energy team to monitor progress of the O&Ts and the energy performance indicators (EnPIs). The measuring and monitoring list is reviewed at least quarterly. Deficiencies are corrected when noted. The time frame for this stage is normally six months to a year and then ongoing with new O&Ts. With extra effort, this stage can be accelerated to be completed in two months.

The third stage is maintaining. In this stage a management review is conducted and management provides feedback on how the EnMS can be improved. The legal requirements are evaluated for compliance. A self-inspection is performed, and corrective action reports (CARs) or preventive action reports (PARs) are written and resolved. A second-party audit is conducted. The time frame for this stage is normally about three years, with no accelerated implementation.

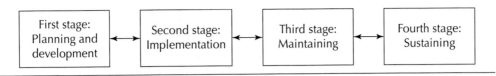

Figure 1.3 The EnMS stages.

The fourth stage is sustaining. The sustaining effort uses fewer resources than the maintaining stage, but every effort is made to meet all the mandatory standard requirements, such as maintaining documentation, evaluating legal requirements, conducting a self-inspection every 12 months, and conducting management reviews at least once a year. The frequency of the energy team meetings could go from monthly to quarterly or twice a year. The number of O&Ts required will probably be reduced. Team members, including the team leader, likely will have changed during the three previous stages, so efforts to keep an active team together will increase. There is not a set time frame for this stage, because it is seen as a journey, not a destination. Thus, it will continue as long as management is committed.

Chapter 2
Requirements

GENERAL REQUIREMENTS
ISO 50001 EnMS Standard

Purpose: To establish the general requirements for an EnMS, such as establish the system, document what you do, implement the system, and maintain and continually improve it in accordance with ISO 50001 EnMS.

General requirements characteristics: Establish the specific boundaries and scope of the EnMS and determine and document how the standard will achieve continual improvement of its energy performance and the EnMS.

Operational Explanation

The general requirements expect an organization to go through all the stages of an EnMS. Define the scope and boundaries of the EnMS so everyone knows what is included and what is not.

The boundaries and scope help focus efforts where needed and are essential for any future audits. The PDCA cycle, shown in Figure 1.2 in Chapter 1, should be reviewed by top management. Following this cycle will ensure that continual improvement is achieved. Top management must commit to fulfilling its roles and responsibilities as spelled out in the standard, and provide leadership and support when and where needed. The EnMS must be planned, implemented, and maintained, and improvements must be continually attained over time. All key elements, documents, and records must be documented and made available to interested personnel in the organization.

Forms, Templates, Processes, and Plans to Meet the Standard

- Write a description of the organization's scope and boundaries
- Use a central documentation system such as Microsoft SharePoint or the organization's IT operations system
- Use the PDCA cycle to continually improve processes and energy performance and document
- Complete all phases and elements with appropriate actions, document, and continually improve

QVS Corporation Example of Implementation

QVS Corporation's boundary is the fence surrounding its property in Gun Barrel City, Texas. This area includes the headquarters building, the distribution center, and the three plants. This information was given in a statement that was included in the minutes of the top management meeting on ISO 50001 EnMS and documented under scope and boundaries in the company's Microsoft SharePoint system for the EnMS. The meeting minutes also show top management's decision to implement ISO 50001 EnMS and its commitment to ensure that implementation is successful. The minutes were documented in the centralized Microsoft SharePoint system under 4.1 General Requirements. The company plans to use the PDCA cycle and follow the EnMS stages shown in Figure 1.3 and in the documented minutes. All the elements should be addressed and documented in the company's central file on Microsoft SharePoint.

MANAGEMENT RESPONSIBILITIES

ISO 50001 EnMS Standard

Purpose: An EnMS cannot be successful without management commitment and action. The standard outlines things that management must do.

Top management characteristics/actions: First, management must establish, communicate, implement, and maintain an energy policy, and it must provide the resources required to develop, implement, maintain, and continually improve the EnMS. Second, management must define the scope and boundaries of the EnMS and communicate to all employees and contractors the importance of energy management. Third, top management must appoint a management representative and allow the establishment of an energy team. The management representative and the energy team will be responsible for the day-to-day energy management activities. They will develop EnPIs and ensure that the results are measured periodically. Top management will incorporate energy considerations in long-range planning to include strategic planning and will perform management reviews. Management reviews can be delegated to the management representative.

Operational Explanation

Without management commitment and support, the EnMS will not be implemented successfully nor will any real improvements in energy performance be realized. Some staff members may try to get the EnMS going on their own, but without any real management support, it will not get off the ground. In implementing ISO 50001 EnMS, members of top management must talk the talk and walk the walk. They must be enthusiastic in the journey; communicate the energy policy, progress, and results; and ask for and receive everyone's involvement in reducing energy use. They must support essential O&Ts and reasonable projects with excellent payback by providing essential resources. They should support management reviews or have their management representative do so. They should ensure measures are in place that show progress and results. Specifically, the head of the organization and top management have the following roles and responsibilities:

1. Develop, implement, and maintain the EnMS
2. Review and approve the essential documents written and issued and decisions made by the EnMS team
3. Provide adequate resources for implementing the EnMS
4. Conduct management reviews
5. Communicate the importance of energy reduction and encourage employees to support efforts to conserve energy
6. Ensure appropriate EnPIs are in place
7. Ensure energy is a consideration in all long-term planning

Forms, Templates, Processes, and Plans to Meet the Standard

- Examples of top management speeches encouraging participation in and support of the energy policy
- Copies of the agenda, minutes, presentations, and sign-in sheets for management reviews
- Central document control system showing that plans, O&Ts, completed records, and live documents are in place
- Approved O&Ts, energy action plans, and projects
- An energy team chartered and functioning with agendas and meeting minutes
- Current EnPIs visible to the organization's people and documented to show results at different time intervals (monthly, quarterly, or semiannually)

QVS Corporation Example of Implementation

The head of QVS delegated the responsibility of appointing a management representative (called the energy champion) to the Strategic Council. The energy champion is responsible for managing the day-to-day EnMS activities, appointing and overseeing the energy team, approving EnPIs, recommending projects to the Strategic Council, and conducting management reviews at least once a year. The Strategic Council established the scope and boundaries for the EnMS when it approved implementation of the EnMS. The scope and boundaries are the headquarters building; Plants A, B, and C; the distribution center (Plant D); and everything inside the fenced area of the complex in Gun Barrel City, Texas. The energy champion ensured that a central document control system was set up on Microsoft SharePoint that contains all EnMS documentation. He had the IT department design the system and put it in place, and then he assigned the energy team, especially the document control officer, to maintain it. All team members and the energy champion have access to any document.

MANAGEMENT REPRESENTATIVE
ISO 50001 EnMS Standard

Purpose: Define the management representative (who has skills and competence) roles, functions, and commitments.

Management representative characteristics/roles: The management representative functions as the energy champion for the organization, ensuring that the EnMS is planned, developed, implemented, maintained, and sustained with continual improvement in accordance with the standard. He or she must identify appropriate personnel to serve on the team or perform energy management activities, along with periodically reporting to management the status, health, and performance of the EnMS. In addition, the energy champion shall ensure planning of energy management activities that are designed to support the organization's energy goals, ensure that the methods used to enable success and control of the EnMS are effective, and continually keep the focus on the energy policy by promoting it throughout the organization. The roles and responsibilities of the members of the energy team and other individuals engaged in energy management functions should be defined, communicated, and praised when earned to ensure an effective EnMS.

Operational Explanation

The energy champion must be respected, able to see the big picture, and able to motivate others and lead a company-wide program to reduce energy and its related costs. He or she should charter a cross-functional corporate energy team and, if needed, plant or facility teams. He or she must keep top management informed. It is good practice to have the team brief the energy champion monthly. The energy champion may want to develop a standard presentation for top management and provide a quarterly update. Any issues that arise between executive updates should be added. Normally, the management representative/energy champion conducts the management review once a year. It is advisable to invite one or more of the top management members to attend the annual management review. The energy champion is the link between the energy team and top management. He or she will present the requests for funds or other resources for projects with good payback periods. The energy champion has the energy team develop EnMS awareness training and send it to all members of management, employees, and contractors at least annually or, if possible, twice a year.

Energy Team

A cross-functional energy team that includes representatives from all the organization's major functional areas will need to be formed by the management representative. The energy champion will need to select a team leader who is responsible, a volunteer, a leader, and interested in energy performance. Sometimes the team members select someone from the team to serve as the team leader. This is okay providing the energy champion approves the selection. A well-respected and experienced facilitator who understands the EnMS should be selected by the energy champion and approved by the energy team leader. Team size should be held to a workable number. The recommended team size is from 6 to 8, but up to 10 is okay if needed to provide proper coverage. The team should meet at least monthly during the planning, developing, and implementing stages.

Forms, Templates, Processes, and Plans to Meet the Standard

- The letter designating the management representative should be signed by the head of the organization or one of the top managers. It should be filed in the centralized document system (Microsoft SharePoint).

- A charter should be developed to spell out the purpose of the team, the responsibilities of team members, frequency of meetings, deliverables, and expectations. The roles and responsibilities of the team may be listed in the charter or included in a corporate procedure for documentation and clarity purposes.

- Microsoft PowerPoint training should be documented in the central document control system.

- All executive updates, including presentation slides and minutes, should be documented.

QVS Corporation Example of Implementation

The QVS Corporation energy team will plan, develop, implement, and maintain ISO 50001 EnMS for the headquarters building, three plants, and the distribution center. The roles and responsibilities of the team were outlined.

Team Leader

1. Work with the facilitator in developing meeting agendas
2. Lead each team meeting
3. Ensure the requirements of ISO 50001 EnMS standard are met
4. Provide the energy champion with a quarterly update on progress and results
5. Ensure O&Ts are developed and implemented
6. Lead the development of the management review inputs and outputs and be the master of ceremonies at the management review
7. Approve the team meeting minutes
8. Represent the team in other meetings and activities where required

Note Taker

1. Take notes during the meeting
2. Serve as a scribe if needed by the team leader
3. Write up the meeting minutes and distribute them in a timely manner to the meeting participants and interested others
4. Ensure that the minutes are filed in the correct file and folder on the facility's computer system

Document Control Manager

1. Maintain documents and records so they are easy to retrieve in the central document control system
2. Ensure that the team uses the most current documents
3. Ensure that the established filing system on the facility's computer system or on Microsoft SharePoint is properly labeled and that the documents or records are in the correct file or folder

Often the team leader volunteers to be the document control manager since he or she is involved with every document produced by the energy team. The author has found this action to be very effective and practical.

Team Members

1. Assist the team leader in achieving the meeting's purpose
2. Assist in achieving O&Ts
3. Identify possible improvements, including new objectives and projects
4. Communicate information about the energy program to the appropriate facility staff

Facilitator

Before the Meeting

1. Make sure arrangements have been made for a meeting place
2. Prepare an agenda that gives the meeting's purpose, location, date and starting time, and items with a time frame for discussion and the responsible person or persons for each, and send to all participants prior to the meeting

During the Meeting

1. Build teamwork
2. Manage conflict
3. Keep team on task and working toward the meeting's purpose
4. Achieve participation

5. Help team reach consensus when possible
6. Evaluate and critique the meeting's effectiveness
7. Ensure that the right tool or technique is used and used correctly
8. Highlight action items and decisions and ensure that the team meets the needs of a decision and its effects before moving forward

After the Meeting

1. Assist any team member in accomplishing his or her task if needed
2. Ensure meeting minutes are accurate, complete, and distributed in a timely manner
3. Assist team leader in briefing management and employees on the status, barriers, issues, and accomplishments

Some teams find it helpful to have an assistant team leader if the team leader has to travel a lot. Often an operations manager, who is appointed to ensure O&Ts and monthly milestones are achieved, serves as an assistant team leader.

The presentations of the energy champion to top management should be included in the documentation system, along with any correspondence.

The energy deployment structure consists of the Strategic Council, the energy champion, the cross-functional energy team, and the facility managers, who also serve as the facility energy managers. Communications flow both up and down the deployment structure.

A BEST PRACTICE: USING OAR AND/OR PAL FOR ENERGY TEAM MEETINGS

What is OAR? What is PAL? What do they have to do with managing meetings? Can they be used for any kind of meeting? These questions and more will be answered in this section.

We have meetings for many different reasons. Staff meetings, department meetings, process improvement team meetings, DMAIC meetings, board of directors meetings, kaizen event meetings, safety meetings, EMS meetings, strategic meetings, scheduling meetings, management reviews, headlight team meetings (used in strategic planning to identify areas of opportunity for developing strategic objectives), quality improvement meetings, and project management meetings are just a few examples of the many different types of meetings. Meetings are how we communicate, coordinate, direct, plan, improve, and gain buy-in on a proposal or new way of doing something. It is hard to imagine a workplace without meetings. In one research study, employees said the biggest reason for them feeling unproductive is the many ineffective and boring meetings they attend.[1] While this section does not directly address the "boring" issue, it does outline how to make each meeting more effective. In doing so, hopefully it will also make the meetings more organized and more interesting, thus mitigating any boredom.

What do OAR and PAL bring to the meeting management arena? Although two different acronyms, they are essentially the same approach. OAR was introduced in 2005 by the author in delivering several quality improvement seminars. PAL has been with us for over 30 years. It was introduced in the early quality management days prior to the Six Sigma era. They stand for:

- OAR: objective agenda restricted

- PAL: purpose agenda limited

Both send the same message. Since PAL is more widely known and accepted, we'll use it. The "P" in PAL simply states that every meeting should have a *purpose*. Write it on the agenda along with the items that will be covered during the meeting. For each item covered, show how much time will be allotted and who is responsible. This step limits the action on each agenda item. Also include in the agenda the title of the meeting, the location of the meeting, and the date and start time of the meeting. The agenda does not have to be long to be effective. An agenda that is one page or less is acceptable and useful. Of course, long meetings with numerous participants could require much longer agendas. Whatever it takes to include the purpose, all agenda items, and all responsible persons is what is needed.

The "limited" part of PAL also means that the meeting is held only for the time needed, and thus most meetings can be completed in an hour or less. Meetings that are not well planned can stretch in time required and frustrate the meeting participants. Let's see how PAL would look for the first meeting of QVS's energy team by reviewing Figure 2.1.

The team leader and the facilitator had prepared some energy policy drafts prior to the meeting and brought them to the meeting to get team members' input. After the meeting (preferably within three days), the note taker or facilitator

---

QVS Energy Team Meeting Agenda: Meeting 1

Date/time: October 20, 2010, 10:00–11:00 AM

Place: Quality Conference Room

Purpose: To review ISO 50001 EnMS, draft an energy policy, analyze utility data, and determine team focus for energy policy deployment.

Agenda:

10:00–10:05 QVS energy team charter	Team leader
10:05–10:15 ISO 50001 EnMS review	Team leader
10:15–10:40 Review several energy policy drafts	Team leader
10:40–11:00 Analyze data and determine our focus	Team facilitator

Figure 2.1 Sample meeting agenda.

should write up the minutes and distribute them to the appropriate personnel, including the objective champion (the designated management representative) and any other interested personnel. This task will be easy if the PAPA (purpose, agenda, points, action items) concept is used.

The PAPA concept works as follows. Locate the agenda in your files and change the file name of "agenda" to "minutes" and save it. Open the saved minutes and change "agenda" in the title line to "minutes." Keep the date, time, location, and purpose as they are. For each agenda item, delete the time and the person responsible but leave the title of the agenda item. Write a summary of the main points that resulted from the discussion. If anyone is assigned to take action on this item, note the action item number and explain who is doing what, when it is being done, and where it is being done, if applicable. At the end of the minutes, give the date and time of the next meeting. The team leader or facilitator may choose to attach any Microsoft PowerPoint presentations or other materials to the minutes. Prior to finalizing the minutes, send them to the team members for their review and see whether they want to add, delete, or change anything. The minutes could look as shown in Figure 2.2.

Using PAPA makes writing meeting minutes easier and ensures that no important information or key points are omitted.

QVS Energy Team Meeting Minutes: Meeting 1

Date/time: October 20, 2010, 10:00–11:00 AM

Place: Quality Conference Room

Purpose: To review ISO 50001 EnMS, draft an energy policy, analyze utility data, and determine team focus for energy policy deployment.

Agenda:

<u>QVS energy team charter:</u> The team leader stated that the Strategic Council has approved a strategic objective "to reduce energy costs." The target, percent reduction from a selected baseline year, will be determined later. The director of operations was selected as this objective's champion. He appointed the vice president of strategic planning as the energy team's team leader and the director of quality and continuous improvement as the facilitator. Eight other team members were appointed to the team, covering all major staff and plant functional areas. Our primary responsibilities are to plan and implement an energy reduction program for our company using ISO 50001 EnMS as a guide.

<u>ISO 50001 EnMS review:</u> The team leader explained the five phases of the ISO 50001 EnMS and the 23 elements that make up the five phases. (See Chapter 1.) He stated that we would use these elements as a guide, including the documentation requirement. The team leader said that he would be the documentation manager for the team and ISO 50001 EnMS.

- Action item #1: The team leader is to set up a file on Microsoft SharePoint for the energy reduction efforts and for ISO 50001 EnMS documentation by elements by December 1, 2010.

Figure 2.2 Sample meeting minutes. *(continued)*

> Review several energy policy drafts: The energy team reviewed drafts of five energy policies that the team leader and the facilitator had found from other companies and modified as appropriate. The team will select an energy policy at the next meeting and send to management for its coordination and approval.
>
> Analyze data and determine our focus: The team analyzed the 2010 energy and water usage cost data. Electricity is 95.17% of the total utility cost and 97.77% of the total energy cost. Therefore, the team decided to focus on electricity cost and kW usage. (See Attachment 2.)
>
> - Action item #2: The team leader will discuss the graphs and team focus with the objective champion by October 28, 2010, and ensure the team is on the right course of action.
>
> Next meeting: The next meeting will be held on November 15, 2010, in the Quality Conference Room starting at 10:00 AM. An agenda will be developed and distributed three days before the meeting.
>
> Approved:
> Team Leader, QVS Energy Team
>
> Attachment #1: ISO 50001 EnMS Five Phases and 23 Elements
> Attachment #2: Utility Costs

Figure 2.2 Sample meeting minutes. *(Continued)*

SUMMARY OF ENERGY TEAM MEETINGS AND MEETING METRICS

Each meeting has a purpose, an agenda, and a limited time frame for the items covered. A person is assigned to each agenda item and is expected to stay within the time frame. It is a goal of the energy champion that at least 80% of the meetings are facilitated and 90% are started and finished on time. After the meeting, the minutes are developed, coordinated with the meeting participants, and filed by year in the organization's EnMS document system. Then the meeting data to be used for calculating the meeting metrics are documented. Meeting metrics are helpful to ensure that the team is on track for achieving its goals. It takes approximately 26 energy team meetings to go from the planning and development stage to the implementing stage and then to the maintaining stage for a normal organization without a guide such as this book. If an organization adopts the required documentation included in this book, the EnMS can easily be implemented in four months, but the results from the team's action plans may take longer (maybe as long as a year after implementation). Table 2.1 reflects a normal implementation. A lot of people prefer not to speed up a complex subject since value may be lost.

Meetings are necessary for any organization to be successful and to continually improve its processes, operations, and performance. Using the tools and techniques described earlier can help your meetings be more interesting, less boring, and more focused and lead to improved effectiveness.

Table 2.1 Summary of energy team meetings and metrics.

Meeting number	Meeting date	Purpose	Start/finish on time? (±5 min.)	Meeting facilitated?	Purpose achieved?	Participants (out of 10)	Deliverables completed?
1	10/20/2010	Review ISO 50001 EnMS and charter	Yes	Yes	Yes	10	N/A
2	11/15/2010	Draft energy policy	Yes	Yes	Yes	9	Draft energy policy
3	12/29/2010	Develop focus, EnPIs	Yes	Yes	Yes	10	EnPIs
4	1/22/2011	Energy profiles	No (+10 min.)	No	Yes	8	Energy profiles
5	2/24/2011	Identify critical success factors (CSFs) and establish O&Ts	No (+30 min.)	Yes	Partially	10	Five O&Ts established
6	3/31/2011	Review O&Ts, finish CSFs and measure, review documentation system	Yes	Yes	Yes	10	CSFs developed. CSF assessment tool used to obtain a score representing their progress. Documentation review completed and no discrepancies found.
7	4/30/2011	Review O&Ts, identify legal and other requirements	Yes	Yes	Yes	9	Legal and other requirements
8	5/24/2011	Review O&Ts, discuss energy audit, identify SEUs, start developing self-inspection checklist	No (+40 min., objective champion attended)	Yes	Partially	10	SEUs identified, others not finished

(continued)

Table 2.1 Summary of energy team meetings and metrics. (*Continued*)

Meeting number	Meeting date	Purpose	Start/finish on time? (±5 min.)	Meeting facilitated?	Purpose achieved?	Participants (out of 10)	Deliverables completed?
9	6/30/2011	Review O&Ts, energy audit planning, develop self-inspection checklist, monitoring and measurement plan, CSF assessment	Yes	Yes	Yes	9	Energy awareness training (sent to everyone but objective champion), including the electricity conservation plan. Three new O&Ts identified and approved.
10	8/30/2011	Review O&Ts, approve energy awareness training, new O&Ts identified (OT-11-06 develop operational controls, OT-11-07 develop contingency plan), develop a self-inspection checklist	Yes	Yes	Yes	9	Energy awareness training (sent to everyone by objective champion), including the electricity conservation plan. Three new O&Ts identified and approved.
11	9/29/2011	Review O&Ts, CSF assessment, review energy audit and selected projects	Yes	Yes	Yes	9	Energy audit and project list
12	10/24/2011	Review O&Ts, project status review	Yes	Yes	Yes	10	N/A
13	11/22/2011	Review O&Ts, develop operational controls for SEUs	Yes	Yes	Yes	7	Operational controls plan
14	12/23/2011	Review O&Ts, CSF assessment	Yes	Yes	Yes	6	CSF assessment
15	1/28/2012	Review O&Ts, project status review, contingency plan	Yes	Yes	Yes	7	QVS Corp. contingency plan for electricity

16	2/26/2012	Review O&Ts, project status review, QVS Corp. procurement plan	Yes	Yes	Yes	9	QVS Corp. procurement plan completed
17	3/30/2012	Review O&Ts, project status review, CSF assessment	Yes	Yes	Yes	9	CSF assessment
18	4/29/2012	Review O&Ts, project status review, possible new O&Ts studied	No	No	No	8	None completed
19	5/30/2012	Review O&Ts, project status review, new O&Ts	Yes	Yes	Yes	9	One new O&T identified
20	6/30/2012	Review O&Ts, project status review, CSF assessment	Yes	Yes	Yes	10	CSF assessment
21	7/29/2012	Review O&Ts, project status review	Yes	Yes	Yes	8	N/A
22	8/26/2012	Review O&Ts, communication plan approved	Yes	Yes	Yes	10	O&T developed. Communication plan completed.
23	9/30/2012	Review O&Ts, finish self-inspection checklist, CSF assessment	Yes	Yes	Yes	9	Self-inspection checklist, CSF assessment
24	10/28/2012	Review O&Ts, project status review, do self-inspection	Yes	Yes	Yes	9	Self-inspection

(continued)

Table 2.1 Summary of energy team meetings and metrics. *(Continued)*

Meeting number	Meeting date	Purpose	Start/finish on time? (±5 min.)	Meeting facilitated?	Purpose achieved?	Participants (out of 10)	Deliverables completed?
25	11/29/2012	Review O&Ts, prepare for management review	Yes	Yes	Yes	8	N/A
26	12/28/2012	Review O&Ts, project review, CSF assessment, management review finalization	Yes	Yes	Yes	5	CSF assessment, management review: Microsoft PowerPoint
Mgt. review	1/25/2013	See agenda		Yes	Yes	10 = objective champion + 2 guests	Approvals and recommendations, new O&Ts

Facts and Metrics

26 meetings from October 20, 2010, to December 28, 2012

22 meetings started and finished on time (metric #1: 84.6%)

24 meetings were facilitated (metric #2: 92.3%)

23 meetings met full achievement of the deliverables (metric #3: 88.5%), 2 met them by partial completion, and only 1 did not meet its deliverables (failure rate is metric #4: 3.8%)

Participation (metric #5): Mean or average is 8.73, median is 9, and mode is 10. Any way you calculate the central tendency for participation, it is good. How about the meeting where only 5 of the 10 participated? Should it be included in the calculation of statistics? It is good for teams to set a participation quorum whereby a meeting counts. Normally, if half of the team is present, the meeting is counted. Some situations are counted providing that at least the team leader and the facilitator were present.

Chapter 3
Energy Policy

CORPORATE VISION FOR ENERGY
ISO 50001 EnMS Standard

Purpose: To have a stated policy approved by management that is compelling and provides direction for the organization in energy reduction, conservation, and other energy actions.

Energy policy characteristics: The energy policy is relevant, is properly scaled to the organization's size, and provides a framework for setting O&Ts. It contains a review process to ensure that the organization is on track, that the system is reviewed regularly and updated as needed, and that the policy is documented and communicated to all levels of the organization. The energy policy should include commitments that resources for achieving O&Ts will be available and that all legal and other requirements will be complied with. The energy policy should encourage the purchase and use of energy-efficient products and services and support designs for energy performance improvement. Finally, the energy policy should commit to continual improvement in energy performance.

Operational Explanation

What Is an Energy Policy?

An *energy policy* is a statement of an organization's policy for managing energy. It is similar to a vision statement for energy. It should be relevant, actionable, long range, easy to understand, compelling, and consistent with policies of other management systems in the organization. Once developed and approved, it should be communicated to all management, supervisors, employees, and contractors. The ISO 50001 EnMS standard requires that the energy policy address all energy used in the defined scope of the EnMS. This includes electricity, natural gas, steam, solar, and other energy sources. The policy should show the organization's commitment to meeting the ISO 50001 EnMS standard requirements, willingness to satisfy all legal and other requirements, and pledge to continually improve energy performance. Increasing renewable energy is a goal of all government and military facilities. It is also a goal of some private companies. For these companies, then, a commitment to increasing renewable energy belongs in the energy policy. The energy policy is a primary driver for implementing and improving your EnMS and energy performance. The energy policy may form part of a wider environmental policy or another policy and could include other commitments such as life cycle costing or total productive maintenance. The energy policy is the key driver in improving your EnMS and energy performance in that O&Ts and energy action plans are developed to ensure the energy policy is achieved.[1]

How Is the Energy Policy Used?

The energy policy is communicated to all employees and contractors. It is important that it is understood by all since everybody has a role and responsibilities that support energy reduction and performance. The energy policy is not company sensitive since it must be available for the public (if top management approves for a private organization), if requested. Customers, public authorities, investors, and others may have a need to review it. Most important is that the legal and other requirements and the energy policy are the primary considerations (provides a framework) in establishing energy O&Ts. It is essential that the energy policy be regularly reviewed, revised when necessary, communicated when revised, and documented in the company's EnMS documents and records filing system.

Forms, Templates, Processes, and Plans to Meet the Standard

- The written energy policy.
- A copy of the evaluation of the energy policy (both the evaluation and the energy policy are placed in the centralized documentation system).
- A copy of the energy policy communication to all employees and contractors (placed in the documentation control system).
- Executive orders for government organizations require renewable energy to be obtained to meet goals. Renewable energy is getting easier to obtain with the increase in wind turbines (1.5 megawatts each that can provide electricity for 300 homes), solar panels on houses, solar plants, more biomass, and geothermal and water power. Green Mountain Energy Company of Austin, Texas, provides competitive electricity rates in Texas while providing 100% pollution-free electricity. It has expanded into New York and now supplies electricity to the Empire State Building, the largest single user of electricity in New York City.

The energy policy should be written by either the energy team or the energy champion. Normally, the energy team develops a draft and the energy champion or his or her designee fine-tunes it. Then the energy champion presents the policy to top management for approval. Therefore, the processes necessary are determining what should be included in the energy policy, developing a draft, reviewing for sufficiency, fine-tuning the draft, and obtaining energy champion and top management approval.

The energy policy will differ from organization to organization but should have some common commitments. The organization will commit to the products or services produced, to the legal and other requirements, to an energy conservation program with all employees and contractors involved, and to continual improvement of the EnMS and energy performance. An example of a commitment statement is shown in Figure 3.1.

QVS Corporation Example of Implementation

The sample energy policy shown in Figure 3.1 has the basic requirements from ISO 50001 EnMS-2011. It should be added to where necessary and put in the format of QVS's other policies, such as those for safety and the environment. Let's look at the draft of QVS Corporation's energy policy as shown in Figure 3.2 and see whether the energy team and the energy champion would recommend their top management to approve and communicate it to their personnel.

We will use an evaluation technique where we establish appropriate criteria and then evaluate the energy policy against them (see Table 3.1). The energy policy must be relevant, easy to understand, actionable, long range, compelling, and congruent with other policies of the organization, and include a phrase about the organization's future intention of obtaining renewable energy.

Our answer would be yes, the energy team would recommend that top management approve the energy policy statement providing it is similar in format to other existing policies. QVS Corporation's top management approved the energy policy presented to them by the energy champion.

We, [Org.'s title], commit to a long-term reduction of our energy consumption and to the improvement of our energy efficiency by using an Energy Management System (EnMS).

We commit our EnMS to meet the requirement standards in ISO 50001 Energy Management System (EnMS)-2011.

We commit to increasing our percentage of renewal energy in the long term.

We will use energy performance indicators (EnPIs) to evaluate our progress and results on a regular, periodic basis and strive for continuous improvement by implementing objectives and targets with well-developed action plans. We will revise the energy policy when needed and communicate it to all our personnel.

Figure 3.1 Sample energy policy.

QVS Corp. is committed to purchasing and using energy in the most efficient, cost-effective, and environmentally responsible manner possible. Therefore, QVS Corp. shall:
- Practice energy conservation at all its facilities
- Lower its peak demand at facilities
- Improve energy efficiency while maintaining a safe and comfortable work environment
- Lower its kilowatt hours per square foot to best-in-class levels
- Increase its percentage of renewable energy used
- Continually improve its performance

Figure 3.2 QVS Corporation energy team's draft energy policy.

Table 3.1 Evaluation/check of the energy policy.

Criterion	Criterion met? (yes/no)	Remarks
Relevant	Yes	Pertains to energy management and performance
Easy to understand	Yes	
Actionable	Yes	Sends a clear message
Long range	Yes	No time period given
Compelling	Yes	Gives energy team a lot of possibilities for objectives and projects
Congruent with other policies	Don't know	Don't have other policies to review, would ask to see them
Includes renewable energy	Yes	

Chapter 4
Energy Planning

AN OVERVIEW

The energy planning process has eight parts:

1. Identifying the scope of the EnMS and understanding the legal and other requirements
2. Understanding past and present energy consumption (determining how much was used and how it was used)
3. Identifying the SEUs
4. Identifying energy-efficient measures that would reduce the energy consumption of these SEUs
5. Selecting a baseline so improvements can be compared in the future
6. Identifying EnPIs for the organization and for the SEUs if possible
7. Developing O&Ts or projects with reasonable and acceptable payback
8. Developing action plans to implement the O&Ts and achieve the desired results

Steps 1–8 are the energy planning process. Steps 2–8 are the energy review. The energy review should be maintained, documented, and reviewed anytime a major energy user changes or the plant or facility undergoes a major modification. If neither of these occurs, then after three years a review should be undertaken and present components validated.

The scope for QVS Corporation was identified previously. Identification of the scope is necessary to ensure the energy policy and O&Ts developed are within the boundaries established for the EnMS. Does the scope include the yard around the facilities with the security lights, the guard house, and, most important, any operations away from the facility? QVS Corporation established its scope to include the five facilities (the headquarters building, Plants A–C, and the distribution center [Plant D]), the yards around the facilities, and one guard house. There are no operations outside the QVS Corporation complex in Gun Barrel City, Texas, covered in the scope.

GENERAL REQUIREMENTS

ISO 50001 EnMS Standard

Purpose: The company, business unit, or organization shall conduct an energy planning process and document and use the results to improve energy performance.

General planning characteristics: Energy planning consists of performing an energy review and developing an energy profile and baseline to measure future results against.

LEGAL AND OTHER REQUIREMENTS

ISO 50001 EnMS Standard

Purpose: To identify the legal and other requirements that relate to energy purchase, use, consumption, and efficiency. These requirements should be available to the organization to review periodically and to ensure that it is in compliance.

Legal requirements characteristics: The organization, business unit, or company should determine how these requirements apply in planning, implementing, and maintaining an EnMS.

Operational Explanation

Any organization that uses energy and wants to implement ISO 50001 EnMS must identify any federal law, state statute, or county or city ordinance that regulates, controls, or addresses energy use or installation and maintenance of energy-providing equipment. The other requirements can be in the form of the organization's policy or procedure or executive order (if the facility is owned or operated by a federal organization). Legal and other requirements should be evaluated every year to ensure the organization is in full compliance. If it is not, corrective action must be taken and managed until the correction has been put in place and the organization is verified to be in compliance.

Forms, Templates, Processes, and Plans to Meet the Standard

The requirement is to research and develop a list of legal and other requirements. It is advisable to separate them into meaningful categories such as purchase, use or consumption, building codes, federal regulations, state statutes, and county and city ordinances.

QVS Corporation Example of Implementation

The legal and other requirements will be the responsibility of the facility managers at QVS Corporation's five facilities and the corporate energy team will evaluate the facilities' compliance during its annual self-inspection of the EnMS. The requirements should be listed in a spreadsheet, in a table, or as shown in Figure 4.1.

The purchase of electricity is determined by utility rates and considerations such as cents per kilowatt-hour (kWh) and other fees and adjustments such as power factor adjustment. This will be spelled out in the agreement with the utility provider and is normally included in the electric bill. It may also be impacted by Texas Senate Bill 7, passed January 1, 2002, regarding electricity deregulation in Texas.

Design and operations are determined by consideration of the following standards and legislation:

- ASHRAE (American Society of Heating, Refrigerating, and Air-Conditioning Engineers) standards for heating, air-conditioning, ventilation, and data centers (3, 15, 62, and 90.1), consisting of three types of standards:
    1. Methods of testing
    2. Standard design
    3. Standard practices
- ASME (American Society of Mechanical Engineers) and IEEE (Institute of Electrical and Electronics Engineers) standards on insulation, high-voltage testing, and hooking items to electronic equipment
- Federal legislation including the National Environmental Policy Act (NEPA), the Natural Gas Policy Act (NGPA), and the Public Utility Regulatory Policies Act (PURPA)
- Electricity at Work regulations that require companies to maintain all electrical items and systems in a safe and workable condition
- Environmental Protection Agency (EPA) regulations for meeting ventilation requirements in offices
- County and community statutes and codes for building installation to include electricity and natural gas requirements
- State requirements for power factor and other issues

Compliance Evaluation

Compliance evaluation of the legal and other requirements will be added to the QVS Corp.'s internal audit or self-inspection checklist and evaluated as part of the annual self-inspection. Documentation of the results, including any corrective action efforts will be maintained.

1. Has each facility been evaluated for electricity legal requirements and are they documented?   ☐ Yes   ☐ No
2. Have specific energy laws been identified and documented?   ☐ Yes   ☐ No
3. Have the legal requirements been reviewed within the past year?   ☐ Yes   ☐ No

    (Date of last review: _____)

4. Is there evidence that the facility has evaluated its compliance with the electricity and natural gas legal requirements?   ☐ Yes   ☐ No
5. If the facility has aspects that are not in compliance with applicable legal requirements, have corrective actions been taken?   ☐ Yes   ☐ No

Comments: _____
_____
_____
_____

Figure 4.1   QVS Corporation sample self-inspection checklist.

ENERGY REVIEW

ISO 50001 EnMS Standard

Purpose: The organization, business unit, or company shall perform an energy review, document it, and maintain and update it periodically. SEUs should be identified.

Energy review characteristics: First, identify what energy sources are available. Second, identify how the energy is used and how much is consumed, both present and past. Determine the facilities, processes, systems, and individuals involved in the energy process, and determine what are the SEUs out of the total energy users. Estimate future consumption and help identify the variables that relate to energy use and consumption and especially those of the SEUs. Identify energy efficiencies or opportunities for future improvement or elimination of waste. Revisit periodically.

Operational Explanation

This requirement is for the organization to know how much energy was consumed and for what purpose. This information and data are needed before any O&Ts can be developed. Using annual data of recent years is the best approach to evaluate consumption. Also, having the data by months can show trends and seasonality.

There are certain technical considerations that should be explored in determining the SEUs. The most used criteria to determine the SEUs are which facilities, equipment, and processes consume the most energy. It is recommended that energy variables be identified to determine or estimate energy consumption. Eventually, an organization should be able to predict, with reasonable accuracy, its future energy consumption. Also, consider the energy uses that offer the most potential for energy savings. For example, an organization that uses T12 lights would likely consume less energy by using LED, T8, or T5 lights. Identify personnel who can potentially save energy, such as the facility managers, the data center operators, and the operations and maintenance manager in the distribution center. Determine whether awareness or competency training would help them do a better job in this reduction effort.

SEUs come from processes, buildings, plant equipment, fixtures, and heating, ventilation, and air-conditioning (HVAC); from data centers; and from where organizations transport energy. There are numerous techniques for identifying SEUs. The first one is to walk through the facilities with the facility managers and operators and ask questions and discuss energy uses for equipment and processes. This walkthrough and identification can be broken down into several different areas, such as conducting a lighting survey and identifying areas of infrequent use (e.g., break rooms, restrooms, mechanical rooms, copier rooms, etc.) so that plans for installing occupancy sensors can be made. Another check is to observe whether the automatic sensors that turn an air handler or a chiller on or off are working. It is not uncommon in older facilities to find equipment that runs continuously. Also check the thermostat settings. Don't be surprised if 72° rather than 76° or 78° is the normal setting in the summertime. It is easier to identify and evaluate the users of electricity, natural gas, or other energy forms, such as steam, separately.

Developing a master list of energy-using equipment that includes their rated loads is recommended. Actual loads using submeters, installed temporary meters, spot meters, or kilowatt (kW) meters can be useful. Process maps, pie charts and other graphs, tables, Microsoft Excel spreadsheets, energy models, and energy mapping are also available for use in identifying SEUs and monitoring them. Organizations have several options for identifying SEUs, and most will use a combination of these. Continual improvement of this process will probably be made by first identifying the major broad categories, stratifying from the major items to the lesser items (that are large users of energy), and identifying the variables that are correlated to the SEUs. This will not only improve the SEUs but also grow the master equipment list and increase the organization's accuracy in predicting energy use.

Forms, Templates, Processes, Plans, and Profile(s) to Meet the Standard

Both annualized and monthly data can be used in the energy profile. The use of graphs and tables is an excellent way to present the data. Microsoft Excel spreadsheets enable the graphs to be produced quickly and accurately. Possible sources of data are energy bills, meter and submeter readings, estimated bills when actual bills are not available, and monthly bill analysis of the tariff structure and adjustment fees for power factor and other costs.

Pie charts and column or bar charts are best for capturing, displaying, and analyzing energy data. Microsoft Excel or a similar system is excellent for creating these charts.

The first cut at the SEUs—focusing on the traditional categories, identifying the possible variables that impact the energy users, and conducting energy audits and/or walkthroughs by facility managers and their maintenance personnel—will lead to an ever-improving list of SEUs. The submetering and the installation and use of other meters should closely follow this evolving list of SEUs. Operational controls should be added where needed so that all SEUs have an O&T to improve their usage or an operational control to minimize their impact or both. SEUs, therefore, should:

1. Be measured either by a submeter, a spot meter, or a kW meter. The latter two should be accomplished annually.

2. Have energy variables identified and then evaluated as to their effectiveness in either normalizing or predicting performance.

3. Have either an O&T or an operational control to mitigate the impact or both.

The recommended process, as shown in Figure 4.2, is:

1. Identify the major energy users

2. Stratify to the SEUs

3. Identify the next lower component or supporting equipment associated with the SEU

4. Select the SEU and show whether it is measured now

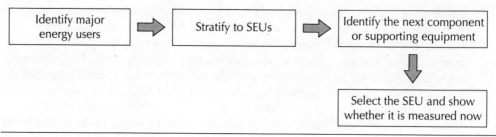

Figure 4.2 SEU selection process.

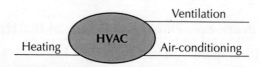

Figure 4.3 Example of stratifying: HVAC.

Start with a major user, such as lighting. Then ask, "What does this unit include?" It includes office lighting, plant lighting, security lighting, exit signs, and occupancy sensors. This is how you stratify. Keep asking this same question until you cannot break out any further. In this case, one stratum is sufficient. However, sometimes more than one stratum is needed. For example, office lighting could include different types of lights such as T12s, T8s, T5s, and LED. These would be the second-level stratum. A third-level stratum could be whether the lights have electronic or nonelectronic ballasts. Keep stratifying until you can go no further. Figure 4.3 shows an example of the stratifying process. At the final stratum, determine what will be measured. Then it can be determined whether additional measurements would be beneficial, such as measurements of the different kinds of lights or of the electronic ballasts versus the nonelectronic ballasts.

The energy review should be accomplished by following the process identified in Figure 4.4.

QVS Corporation Example of Implementation

By following the process identified in Figure 4.4, we develop an energy profile. Let's understand QVS Corporation's energy profile. QVS established a corporate goal of reducing electricity by 10% from the 2010 baseline by 2015 and another 10% by 2017. Table 4.1 shows the electricity intensity and the contribution needed by each facility to meet the 10% goal.

The energy team decided to research the energy and water utility costs for 2010. They are shown in Figure 4.5.

The cost of electricity is 95.17% of the total utility cost and 97.77% of the total energy cost. Therefore, the team decided to focus on electricity cost and kilowatt-hour (kWh) usage for now. It is apparent from Figure 4.5 that electricity is the low-hanging fruit on the tree or the big rabbit to catch first and address for reduction.

ENERGY PLANNING   31

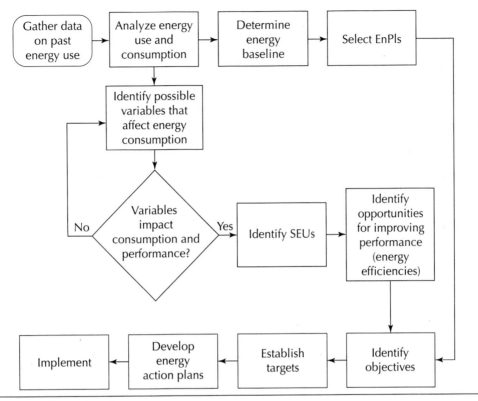

Figure 4.4 Example review.

Table 4.1 QVS Corporation 2010 electricity profile.

Facility	kWh usage	Gross square footage	Electricity intensity (kWh/sq. ft.)	Cost (dollars)	Targeted contribution to achieve target 10% reduction (kWh)
HQ	2,681,740	55,000	48.76	174,313.10	268,174.0
Plant A	3,495,709	102,500	34.10	227,221.08	349,570.9
Plant B	3,423,075	125,000	27.38	222,499.87	342,307.5
Plant C	3,275,166	98,000	33.42	212,885.79	327,516.6
Plant D	3,245,210	100,000	32.45	210,938.65	324,521.0
QVS total	16,120,900	480,500	33.55	1,047,858.49	1,612,090.0

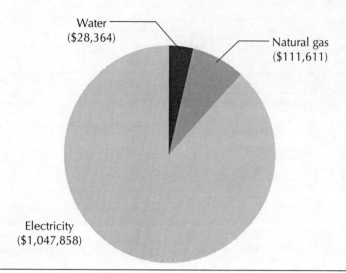

Figure 4.5 QVS Corporation 2010 utility costs.

Developing QVS Corporation's Electricity Profile

The steam energy source was converted to natural gas. The kWh usage data were discussed with the champion and presented to the Strategic Council. They established the following goal: "QVS Corporation will reduce electricity consumption by 10% from calendar year 2010 baseline kWh usage by the end of 2015, with an additional 10% by 2017."

The natural gas usage for 2011 should be approximately $111,611 or 20,074 cubic feet. The present cost is $5.56 per cubic foot. Figures 4.6–4.10 show the kW

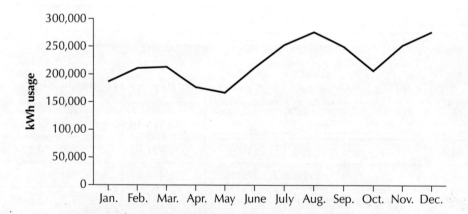

Figure 4.6 QVS Corporation HQ facility 2010 electricity profile.

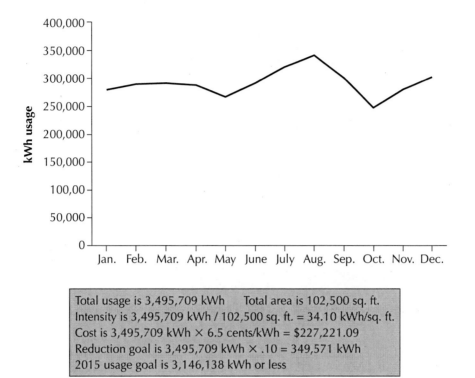

Figure 4.7 QVS Corporation Plant A 2010 electricity profile.

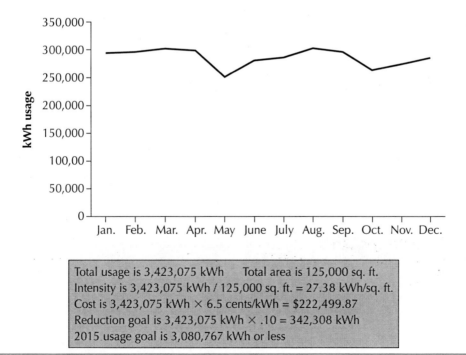

Figure 4.8 QVS Corporation Plant B 2010 electricity profile.

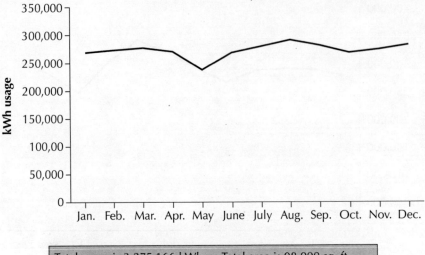

Figure 4.9 QVS Corporation Plant C 2010 electricity profile.

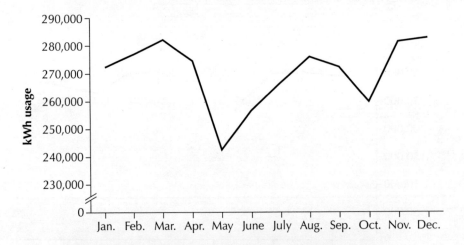

Figure 4.10 QVS Corporation Plant D 2010 electricity profile.

usage in 2010, and the cost and electricity intensity for each facility in the EnMS are outlined.

These data represent the basics for an energy profile required in ISO 50001 EnMS paragraph 4.4. The energy profile should also include similar charts and information for natural gas. Natural gas consumption will be added to the electricity consumption to produce EnPIs. Both kWh for electricity and cubic feet for natural gas will need to be converted to a common unit, the British thermal unit (Btu). This conversion will be shown later when EnPIs are calculated.

Identifying SEUs can be accomplished using the process previously shown. The process is as follows:

1. Identify the major energy users (see first column in Table 4.2)

Table 4.2 Stratifying SEUs.

Major energy user	SEU	First stratification	Second stratification	Selected SEU/ measured?
HVAC	Heating	Boiler	Boiler components	Boiler/not measured
		Furnace	Furnace parts	Furnace/not measured
	Ventilation	Ducts		
		Roof air intake equipment		
	Air-conditioning	Air conditioner/ computer room air conditioner (CRAC)	Chillers	Chillers/not measured/ air conditioners and CRACs selected
			Air handling units	
			Coils	
			Economizer	Economizer/not measured
Lighting	T12s	Lamps		Lighting/not measured
		Ballasts		
	T8s	Lamps		
		Ballasts		
	T5s	Lamps		
		Ballasts		
	Exit signs			
	Security and parking lights			

2. Stratify to the SEUs (see the second column of Table 4.2)

3. Identify the next lower component or supporting equipment associated with the SEU (see the third and fourth columns of Table 4.2)

4. Select the SEU and show whether it is measured now (see the fifth column of Table 4.2)

Sometimes it may not be necessary to have a major user. Just start at the SEU, then stratify to another component or to equipment that supports the possible SEU.

The final SEUs for QVS Corporation are shown in Table 4.3.

Table 4.3 QVS Corporation's SEUs.

Major energy user	SEU
HVAC	
Heating	Boiler
Air-conditioning	Air conditioner
Air-conditioning	Chillers
Air-conditioning—data center	CRACs
Air-conditioning	Economizers
Ventilation	Roof ventilators
Lighting	Office lights
	Plant lights
	Security lights
	Exit lights
Office machines	Computers, monitors, and laptops
	Imaging machines (copiers, printers, and fax machines)
	TVs and other electronics
Motors, machines, shop equipment, automated distribution center	

Source: "Maintenance Checklist," Energy Star, http://www.energystar.gov/index.cfm?c=heat_cool.pr_maintenance.

IDENTIFYING ENERGY EFFICIENCIES

Heating, Ventilation, and Air-Conditioning (HVAC)

Climate control is primarily provided through a central heating, ventilation, and air-conditioning (HVAC) system, often a packaged unit that contains more than one function. The HVAC system regulates several essential variables, such as temperature, humidity, and indoor air quality. First, the HVAC draws outdoor air into the air handling subsystem and then conditions this air in either the chiller (cooling) or heating subsystems to the correct temperature and humidity. Next, the system transfers or vents the conditioned air to locations throughout the building as needed. Improving the HVAC system depends on the particular characteristics of the facility or building, the condition of the existing equipment, and any previous problems with indoor air quality.

Central HVAC systems are usually oversized and are not set for optimal energy efficiency. Before adding additional cooling capacity for your facility, have a mechanical engineer inspect the system to see whether a redesign or revamp may help it to be more efficient, thus saving considerable dollars. Several companies have saved as much as 40%–50% on their electricity consumption by adjusting control set points and modifying existing equipment. Optimizing HVAC equipment also lowers CO_2 emissions, thus improving the environment. In fact, saving 10 kWh of electricity will save 7.3 pounds of greenhouse gas (GHG) emissions.[1]

To save additional funds, consider installing an enhanced automated control system or building automation system (BAS), along with compatible control equipment on building mechanical and electrical systems. Most government buildings have BASs, but a lot of them need updating. The HVAC, lighting, fire and smoke detection, advanced meters (electrical, natural gas, and water), and security systems can all be integrated into an automated control system. Enhanced automation allows facility staff greater zone control by continuously monitoring and adjusting lighting and HVAC equipment based on people densities, the environment, and other factors. This can also allow for load curtailment during peak times when utility rates are highest. Enhanced automation can decrease employee complaints arising from poor indoor air quality since zones rarely fall outside control set points. Problems that do arise are detected quickly and fixed, often before the employees even notice. Because equipment is optimized and operated in coordination with the full system, equipment life is extended and downtime for repair (mean time to repair) is minimized. Regular use of a BAS can save 5%–7% of your electricity use.[2]

Before undertaking major costly upgrades or retrofits, investigate ways to reduce heating and cooling loads. Lighting upgrades and building envelope improvements such as adding insulation, installing energy-saving windows, and using energy-efficient office equipment such as Energy Star all reduce heat buildup and lower cooling loads. Try these energy-reducing methods first, and make HVAC upgrades only if nothing else works. This allows engineers to size the HVAC system, minimizing overall costs while ensuring that the entire building operates efficiently and comfortably.

Lighting

Lighting represents approximately 21% of all electricity consumed in commercial office buildings[3] such as the headquarters building of the QVS Corporation. Thus, lighting is a major part of total building operating costs. Retrofits (in older buildings) to lighting systems can yield savings as high as 30% of current costs and can easily be incorporated into the building's preventive or general maintenance program.[4]

When retrofitting a lighting system, one must consider two related system components: the ballast and controls. Replacing magnetic ballasts with basic electronic ballasts can save a minimum of 12% of energy consumption. Premium ballasts come in three main types: instant-start, program-start, and dimmable. Controls, which should be matched to the appropriate ballast, greatly improve efficiency while offering greater flexibility to the user.[5] Occupancy controls shut off lights in empty areas, and photo sensor controls dim or shut off lights when natural light renders them unnecessary. Timers may be used with both types of control systems. Controls are very popular in organizations' electricity conservation programs. Don't overlook using them; they are practical and will reduce your energy use.

Improvements in fluorescent luminaries have reduced energy consumption while improving light quality. Switching out T12 fluorescents for efficient T5s results in higher efficacy (i.e., fewer watts per lumen). T5s last longer and require less maintenance over the life of the lamp. The light quality is increased; thus, fewer fixtures per square foot are needed. Fewer watts per fixture along with fewer fixtures per square foot results in lower cooling loads. This means that related cooling costs are also reduced.

All lights have a lamp of some sort. The fixture connects and positions this lamp to its energy source—electricity. In fluorescent lights, the energy supply is modulated through ballasts. Halogen and incandescent fixtures, on the other hand, connect directly to the energy supply. The lamp, ballast, and fixture are called a "luminaire." Light is the product of a luminaire.[6]

Cleaning light bulbs and fixtures to increase lighting output levels that have been reduced by dirt and dust can save as much as 10% of electricity consumption.[7]

Office Machines

Even when they are turned off, office machines that are plugged into an electrical outlet still use electricity. Computers, monitors, and laptops can easily be set to go to sleep during idle periods if the organization implements IT power management. For monitors, engage the Energy Star "sleep" feature. For computers and laptops, set the system to either hibernate or standby when the machine has been idle for around 15–20 minutes. This is an easy way to reduce electricity use without decreasing productivity or hindering computer operations.

Procurement should purchase only Energy Star or energy performance environmental attributes tool (EPEAT) (government) computers, monitors, laptops, copiers, printers, fax machines, and televisions. Specifications should be written to include this requirement. An organization should develop and communicate a policy or procedure for purchasing Energy Star or EPEAT items if available and include (or do separately) a policy or procedure for implementing IT power management. Free software is available that can check the network to ensure power management features are set to turn off the computers and monitors when not being used.

An organization can save 20%–30% of office paper (fewer kWh needed) by implementing an office paper-reduction program.[8] The primary focus of this program is to use more electronic files and do duplexing (two sides). The list will quickly grow as you identify other ways to save paper.

Motors

Look for motors that are operating unnecessarily and shut them down. This could be as simple as a ceiling fan running in an unoccupied space or as complicated as cooling tower fans still running after target temperatures have been reached. Both of these situations are common in manufacturing facilities.

Mechanical problems are the main cause of electric motor failures. Routine maintenance such as lubricating bearings, and checking for adequate and clean ventilation is important. Also, ensuring that motors do not experience any voltage imbalance will help them achieve their full-life potential while minimizing their energy consumption.

Air Compressors

Check hoses and valves for leaks regularly and make any needed repairs. An improperly maintained system can waste between 25% and 35% of its air due to leaks. Cleaning intake vents, air filters, and heat exchangers periodically will increase both equipment life and its efficiency.[9]

Boilers

Your preventive maintenance program should include a work order for treating makeup water to prevent equipment damage and efficiency losses. If not, buildup inside the tank could result and decrease heat transfer to the water. This would necessitate more frequent blow down, wasting both water and energy. Also, be sure to check the air to fuel ratio to ensure that the combustion process is operating efficiently. This will enable excellent combustion efficiency.[10]

Fans, Bearings, and Belts

Fan blades, bearings, and belts should be inspected at least annually to prevent failure. The fan blades should be cleaned. The bearings should be examined for adequate lubrication. Belts should be adjusted or changed if needed.

Building Envelope and Seals

One major source of energy loss is air infiltration (air leaking out through gaps around doors on the receiving and loading docks or any doors entering the facility). At least once a year, check and repair gaps in door seals. Emphasize to all employees to keep the doors closed. Check the roof attic and any outside walls to see whether insulation should be added to help reduce energy use.

Air Filters

Change air filters every one to three months. Air conditioners that use an economizer and that are located next to highways or construction sites will need more frequent filter changes.

Leaks

Cabinet panels and ducts on rooftop HVAC equipment should be checked quarterly for leaks. A structural check should be made to ensure that the units are secure (with all screws in place). Every year, inspect all access panels and gaskets (particularly on the supply-air side, where pressure is higher). One leak in an HVAC rooftop unit can cost $100 per unit per year in wasted energy.[11]

Condenser Coils

Check a unit's condenser coil quarterly to see whether it needs to be cleaned, and clean it annually to get rid of any debris. A dirty coil that raises condensing temperatures by 10° Fahrenheit can increase power consumption by 10%. This results in about $120 in electricity costs for a 10-ton unit operating 1000 hours per year.[12]

LEAN ENERGY ANALYSIS

The lean energy analysis (LEA) process of identifying energy-saving opportunities by considering that "any energy use that does not add value to the product or plant environment is waste" is useful.[13] The LEA statistical methodology uses as few as 48 data points, which are obtainable through an on-site data collection effort and interviews with facility management to validate the data findings. This analysis enables multivariable change-point models of electricity and natural gas usage as functions of outdoor air temperature and the quantity of production. The statistical models are used to subdivide plant energy use into facility, space-conditioning, and production-related components.[14] The Air Force Center for Engineering and the Environment (AFCEE) claims the results of the LEA provide a quick and economical method of bracketing energy-efficiency potential in a facility, measuring control and direct energy conversion efficiencies. This method may have great potential for EnMS users in the future.

There are other energy-efficiency methods or processes that will reduce SEU consumption:

1. *IT power management:* Enabling the power management features for monitors (Energy Star sleep function) and computers and laptops (hibernate or system standby) will save a lot of kWh and approximately $75 per year for each one. A monitor or computer that has these features enabled will automatically use much less power after it has been inactive for 20 minutes.

2. *Thermostat settings:* Setting the temperature at a less power-consuming level, while still keeping the employees comfortable, is another way to save energy. To do so, the organization may have to implement strategies to improve the climate and culture so that management, employees, and contractors will accept changes without major complaints or a decrease in morale. Sometimes asking them to wear an extra layer of clothing in the winter may be sufficient to reduce most of the complaints. The Energy Saver website of the Department of Energy (DOE) states that reducing 1° on the thermostat during the winter can save 3% on the electric bill for heating.

3. *Efficient lighting:* Installing LED and T5 lights is an energy-efficiency measure.

4. *Occupancy sensors:* Installing occupancy sensors in less used areas is an energy-efficiency measure. Mechanical rooms, break rooms, hallways, imaging rooms, restrooms, and so forth, are excellent candidates for occupancy sensors.

5. *Data centers:* In data centers, separating the servers into hot and cold aisles is an energy-efficiency measure. Additionally, separating the hot air from the cold air on the way back to the computer room air conditioner (CRAC) can save up to 58% of electricity consumption and cost.[15] If the data center is cold, the organization is using too much energy. Also, consult the recommended temperature ranges in the appropriate ASHRAE (American Society of Heating, Refrigeration, and Air-Conditioning Engineers) standards and adjust temperatures to stay in range but higher than presently set if the room is cold.

6. *Duplex printing:* Implementing an office paper-reduction program by emphasizing the use of electronic files and duplex printing can save from 20% to 30% of present paper use and cost and reduce electricity use for imaging equipment.

7. *Day lighting:* Incorporating more daylight into new office facilities is an excellent energy-efficiency measure.

8. *EnMS:* If building temperatures are not controlled by an EnMS or a BAS, a programmable thermostat can be used to increase energy savings and enhance comfort by automatically adjusting to preset levels. It can also adjust temperatures on weekends and holidays.

9. *Economizers:* Many air-conditioning systems use a dampered vent called an economizer to draw in cool outside air when it is available, reducing the need for mechanically cooled air. The linkage on the damper, however, can seize up or break if not regularly checked. An economizer that's stuck in the fully open position can add as much as 50% to a building's annual energy bill by allowing hot air in during the air-conditioning season and cold air in during the heating season. Have a licensed technician calibrate the controls; check, clean, and lubricate your economizer's linkage at least once a year; and make repairs if necessary.

10. *Plug loads:* Small items such as computer speakers, radios, and coffee makers can burn a significant amount of energy. Like computers, office equipment such as printers and fax machines also have energy-saving settings. Procurement may require that any equipment purchased have these settings. Smart strips sense when devices are in "off" mode and will cut all power to the devices plugged into them, eliminating so-called phantom loads. Smart strips that control loads based on occupancy are also available and a wise choice. Power strips should be given to all employees, giving them an easy way to switch off all their often-forgotten energy users at the end of the day. Plug-load reduction is an excellent conservation action.

11. *Space heaters:* Space heaters are energy hogs, drawing one kilowatt or more of power per hour. It is energy efficient to plug heaters into power strips controlled by occupancy sensors (other loads, such as task lights and monitors, can also be plugged into power strips). Recognize that the perceived need for individual space heating usually signals poor HVAC

system control. Also, space heaters can be a safety hazard. Use only if you have no other alternative.

12. *Outside-air intake controls.* Many facilities have rooftop units for heating, ventilation, and sometimes cooling. These are often equipped with exhaust fans that bring in outside air for ventilation. Although the units should be set to run only when spaces are occupied, it is not uncommon to see them run 24 hours/7 days a week. This can be a real waste of electricity.

13. *Vending machine controls:* Use occupancy sensors to power down vending machines when the area is unoccupied for a long period of time. Sensors can save nearly 50% of the $150 annual electricity costs for a single vending machine.[16]

14. *Regular or preventive maintenance and cleaning:* Maintaining the facility and equipment is important, both to save energy and to protect equipment.

15. *Process heating:* There are several ways to improve energy efficiency in process heating. Optimizing the ratio of air to fuel with flow metering or flue-gas analysis is a best practice to maximize burner efficiency. For indirect heating systems, inspect and clean heat-transfer surfaces periodically to avoid soot, scale, sludge, or other buildup that can significantly reduce system efficiency. Reduce air infiltration into the heating process by repairing system leaks and keeping furnace doors closed. Process heating can be an SEU.

In addition to the LEA, base load analyses and energy audits using energy experts who conduct an EISA (Energy Independence and Security Act) type of audit that includes identified improvements but also shows a payback period are excellent methods for identifying energy efficiencies. Using Six Sigma, metering reviews, and plans such as installing advanced meters including submeters; identifying and adopting best practices and/or best available techniques (BATs); installing predictive maintenance or reliability centered maintenance; and performing more root cause analysis, Pareto analysis, and benchmarking are all possible techniques for identifying energy-efficiency measures. An example of lean analysis comes from the author, who was checking energy use in a facility. In doing so, he ran across a container that was filled with warm water. When he asked what it was used for, he was told that it was for washing parts in a former process that was changed and moved to another part of the facility about six months earlier. Every day, this water container was heated automatically and no one questioned the purpose.

Energy monitoring is a must for improving energy performance. Employee suggestions could prove very beneficial. Implementing the 5 Ss, a tool to achieve orderliness and cleanliness that helps improve employee efficiency and productivity, could prove to be a good energy-efficiency measure. Utilities publish energy-reduction tips that need to be studied as to possibly implementing in your organization.

Once the energy auditors have completed a walkthrough of the facility, it will be helpful to list the energy efficiencies described earlier and prioritize them as to the amount of energy savings they can provide for the facility if implemented. For each item selected, identify the cost and the amount of energy that could be saved if the item is installed or fixed.

IDENTIFYING ENERGY VARIABLES

For each SEU, identify the energy variables that are possibly correlated with the SEU or can be used to normalize the data. Table 4.4 shows the first cut at identifying energy variables for the SEUs.

Some of the variables can be measured and analyzed as to their correlation or impact. Often, important equipment that uses a lot of energy cannot be measured since few or no submeters are installed. For example, if the electricity meter in a data center records only the total facility electricity usage, specific power used by the servers cannot be determined (it must be estimated). It would be advisable to install a submeter to measure only the kWh use of the data center. This would not include electricity used by the office managers, system designers, facility maintenance, or other functions.

Why is it important to identify the energy variables? There are several reasons. One is to predict future energy use. Another is for normalization, which enables comparisons to be made or a better understanding of what happened. And yet another reason is to know more about the SEU. If we know that a variable impacts an SEU, we can put an operational control in place to neutralize the effect or at least minimize its impact; some operational controls can do both. For HVAC, the number of people and the amount of square feet are excellent to divide the kWh consumed by. This enables an organization to compare with other similar facilities. Regression analysis could be used to show that when either the number of people or the number of square feet being cooled or heated increases, the kWh consumption will also increase. The number of computers is very sensitive to kWh usage. The more computers an organization has, the more kWh it will consume. Of course, utilizing IT power management could result in a savings of around $75 a year for each computer. The thermostat setting, as we all know, is a variable that, in the summer, the lower the setting, the higher the kWh consumption, and vice versa in the winter. The amount of savings per degree is hard to believe, so why don't we just turn up the thermostat and save all that energy. The reason is that to be comfortable, the temperature of most government buildings and many private ones is set around 72° in summer. To set it at 76°, the organization will likely receive a lot of complaints and experience a loss of employee and contractor productivity. Same for the winter settings. To make it cooler, then (less heat), employees would have to dress differently. The US Air Force set its thermostats at energy-saving temperatures but first instructed its members on why it was important to save energy. Saving energy costs allows more funds for the air force's primary mission of flying and fighting to defend America.

Heating degree days (HDDs) is determined by comparing the present temperature with the temperature on a day in which no heating is required to keep personnel comfortable. In Gun Barrel City, Texas, this temperature is 68° indoors. According to Wikipedia, to calculate HDD, "take the average temperature on any given day, and subtract it from the base temperature. If the value is less than or equal to zero, that day has zero HDD. But if the value is positive, that number represents the number of HDD on that day. . . . HDD can be added over periods to time to provide a rough estimate of seasonal heating requirements." As an example, the number of HDDs for New York City is 5050, whereas that for Los Angeles is 2020—meaning that the houses in Los Angeles can be heated for about

Table 4.4 QVS Corporation's SEU energy variables.

Major energy user	SEU	Energy variables
HVAC		
Heating	Boiler	• Square footage of facility that is heated • Temperature setting on thermostats • Outside air temperature • Heating degree days
Air-conditioning	Air conditioner	• Number of employees • Square footage of facility that is air-conditioned • Temperature setting on thermostats • Percentage of rooms with balanced air • Outside air temperature • Number of days with outside temperature over 100° • Mean temperature • Cooling degree days
Air-conditioning	Chillers	• Number of employees • Square footage of facility that is air-conditioned • Temperature setting on thermostats • Percentage of rooms with balanced air • Outside air temperature • Number of days with outside temperature over 100° • Mean temperature • Cooling degree days
Air-conditioning—data center	CRACs	• Number of servers • Server utilization • Average age of servers • Temperature settings compared with ASHRAE standard • Average temperature maintained • Number of CRACs • Square footage of the data center • Power utilization effectiveness • Cubic feet per minute (CFM) of airflow and temperature at input side of server
Air-conditioning	Economizers	• Outside air temperature • Humidity • Cooling degree days
Ventilation	Roof ventilators	• Outside air temperature • Humidity • Wind

Table 4.4 QVS Corporation's SEU energy variables. *(Continued)*

Major energy user	SEU	Energy variables
Lighting	Office lights	• Working hours • Number of shifts • Number of employees and contractors on site • Number of workstations • Type of lights • Square footage of office space or percentage of total gross square footage that is office space • Percentage of lights backed up by a generator • Ambient light levels
	Plant lights	• Working hours • Number of shifts • Number of employees and contractors on site • Number of workstations • Type of lights • Square footage of office space or percentage of total gross square footage that is office space • Percentage of lights backed up by a generator • Ambient light levels
	Security lights	• Number of shifts • Hours of darkness • Size of parking areas
	Exit lights	• Number of exits
Office machines	Computers, monitors, and laptops	• Number of computers, monitors, and laptops • Percentage of electronics that are Energy Star or EPEAT • Number of office employees and office contract employees • Amount of office paper used • Number of computers, monitors, and laptops with power management features engaged
	Imaging machines (copiers, printers, and fax machines)	• Percentage of electronics that are Energy Star or EPEAT • Number of office employees and office contract employees • Amount of office paper used • Number of copiers, printers, fax machines • Number of electronics with power management features engaged
	TVs and other electronics	• Number of units • Hours used

(continued)

Table 4.4 QVS Corporation's SEU energy variables. *(Continued)*

Major energy user	SEU	Energy variables
Motors, machines, shop equipment, automated distribution center		• RPM • Torque • Horsepower • kW standard versus actual kW • Percentage of time in use • Age of equipment • Mean time between failures (MTBF) • Maintenance schedule • Mean time to failure (MTTF) • Availability • Production quantities • Throughput • Total square footage • kW of refrigeration used • Number of products • Storage capacity • Average time doors are open for loading • Unplanned shutdowns

two-fifths of what it takes to heat the houses in New York City.[17] Gun Barrel City, Texas, falls in between at 3980 HDDs. This metric can be useful in normalizing data for comparing one part of the country or world with another.

Mean temperatures for a region can be found on the internet and can be useful in normalizing. The number of days in which the temperature exceeds 100° is useful in Gun Barrel City but not in Alaska. HDD is useful regardless of location.

Lighting has several variables that impact its kWh usage. One variable is the design; for example, T12 and T5 fluorescent lights use different amounts of kWh. The amount of square feet that the lights need to cover will mostly determine the quality of the light, unless the square footage was a factor in the design and installation of the lights. Fluorescent lights can be stratified into two SEUs: the lamp and the ballast. If the ballast is magnetic rather than electronic, making it electronic will make it energy efficient.

ENERGY BASELINE

ISO 50001 EnMS Standard

Purpose: The organization, business unit, or company will establish a baseline of energy consumption against which future progress can be measured.

Energy baseline characteristics: Changes in energy use will be measured against the baseline. The baseline is normally a calendar year or fiscal year. The energy baseline must be documented and easily accessible.

Operational Explanation

The electricity kWh usage was calendar year 2010. Here we are showing only electricity use, but 2010 is also the baseline year for natural gas. All future comparisons to determine results and performance will be with the 2010 baseline. The normal practice of determining a baseline is to gather two full calendar years of data and then plot in a column graph or line graph. Look for seasonality, constant loads, and the occurrence of any unusual events or disasters. Select as the baseline the most average or representative year.

Forms, Templates, Processes, and Plans to Meet the Standard

- Use the process described earlier and select a year that is considered representative. Both the monthly and the annual cumulative figure will be the baseline (may also be called a benchmark).
- The baseline is the data for each month of a calendar year or fiscal year, depending on the organization's budget time frame.

QVS Corporation Example of Implementation

QVS Corporation's baseline is shown in Table 4.5. Columns 2 and 4 show the baselines. Gross square footage is shown since kWh usage divided by the facility's gross square footage equals the electricity intensity.

EnPIs

ISO 50001 EnMS Standard

Purpose: The organization, business unit, or company should identify and develop EnPIs that measure its overall energy performance.

EnPI characteristics: The EnPIs should be graphed, show trends, be current, and reflect countermeasures or natural events if and when they occur.

Table 4.5 QVS Corporation 2010 electricity baseline for all facilities.

Facility	kWh usage	Gross square footage	Electricity intensity (kWh/sq. ft.)
HQ	2,681,740	55,000	48.76
Plant A	3,495,709	102,500	34.10
Plant B	3,423,075	125,000	27.38
Plant C	3,275,166	98,000	33.42
Plant D	3,245,210	100,000	32.45
QVS total	16,120,900	480,500	33.55

Operational Explanation

The energy savings are equal to the baseline period energy use minus the reporting period energy use plus or minus any adjustments in the baseline or reporting period. The baseline period selected was for the year's cumulative energy use. The amount of energy used in 2012 is subtracted from the baseline (2010) to determine the amount saved or reduced. EnPIs can be set at both management and operational levels. When set by management, they are called corporate measures or indicators. Operations for a department or facility can be set and used to manage its system, which links to the corporate EnPIs. They can have the same indicators and possibly some that pertain only to their facility. For example, only the QVS Corporation's headquarters facility has a data center. QVS would have power utilization effectiveness (PUE) as an EnPI for the data center, which is a SEU for that facility. As the team improves on identifying SEUs, stratifying to lower equipment components, and installing submeters, the EnPIs will increase.

Forms, Templates, Processes, Plans, and Indicators to Meet the Standard

First, the EnPIs are selected. They are shown on a column or bar graph along with a data table showing each month's use.

For each EnPI selected, a data collection plan should be developed and followed. It should include the items shown in Figure 4.11.

The data points should be shown under the graph. Select the unit of measure that is most appropriate for the audience. At the facility, kWh used will be easier to understand than Btu used, but for comparison with other organizations, Btu may be the most appropriate measure.

QVS Corporation Example of Implementation

QVS Corporation's energy use consists of electricity and natural gas. Their units of measure are different, however, so it is inaccurate to just add them. They first have to be converted to the same units. Natural gas bills normally show the cubic feet of natural gas consumed. The electricity units are billed as kWh. It is easy to convert both to Btu. Multiply the kWh by 3412.14 Btu/kWh and the cubic feet by

---

EnPI: _____

Title: _____

Data source: _____

Unit of measure: _____

Frequency of use (circle one):   Monthly   Quarterly   Semiannually   Annually

Person responsible: _____

Date of data gathering: _____

Figure 4.11   Sample data collection plan.

1020 Btu/cu. ft. and then sum. This gives you the total energy consumed. The unit of measure for energy is kBtu/sq. ft., with *k* meaning 1000; dividing by the total gross square footage of the facility gives an energy intensity indicator.

The first EnPI is 2010 kWh/sq. ft. minus 2012 kWh/sq. ft.. This is the electricity intensity (see Figure 4.12). The second EnPI is 2010 cu. ft.(nat. gas)/sq. ft. minus 2012 cu. ft.(nat. gas)/sq. ft. This is the natural gas intensity savings (see Figure 4.13). The third EnPI is 2010 kBtu – 2012 kBtu = energy savings or performance.

EnPI: Savings (baseline 2010 electricity intensity minus current-year electricity intensity)

Title: Electricity Reductions

Data source: Electric bills from each facility

Unit of measure: kWh/sq. ft.

Frequency of use (circle one): Monthly   Quarterly   Semiannually   **(Annually)**

Person responsible: Energy team leader

Date of data gathering: As soon as the December electricity bill is available

(2010 kWh/sq. ft.) – (2012 kWh/sq. ft.)

(16,120,900 kWh/480,500 sq. ft.) – (15,635,029 kWh/480,500 sq. ft.)

33.55 kWh/sq. ft. – 32.54 kWh/sq. ft. = 1.01 kWh/sq. ft.

1.01 kWh/sq. ft. / 33.55 kWh/sq. ft. = 0.03 or **3% reduction**

Figure 4.12 QVS Corporation first EnPI.

EnPI: Natural gas intensity savings (2010 cu. ft./sq. ft. – 2012 cu. ft./sq. ft.)

Title: Natural Gas Intensity Savings or Reduction

Data source: Natural gas bills

Unit of measure: cu. ft./sq. ft.

Frequency of use (circle one): Monthly   Quarterly   Semiannually   **(Annually)**

Person responsible: Energy team leader

Date of data gathering: As soon as the December natural gas bill is available

(2010 cu. ft./sq. ft.) – (2012 cu. ft./sq. ft.)

19,046.25 cu. ft./sq. ft. – 18,105 cu. ft./sq. ft. = 941.25 cu. ft./sq. ft.

941.25 cu. ft./sq. ft. / 19,046.25 cu. ft./sq. ft. = 0.049 or **4.9% reduction**

Figure 4.13 QVS Corporation second EnPI.

It is the corporate EnPI (see Figure 4.14). The fourth corporate EnPI is 2010 kBtu/sq. ft. – 2012 kBtu/sq. ft. = energy intensity (see Figure 4.15).

These four EnPIs are the primary corporate EnPIs. During the implementation, others will be added so the "check" phase can measure SEUs and other important areas and the energy team can have valuable monitoring information. Data center and SEU measurements will make up the majority of the additions. It is wise to establish the big EnPIs first, then add to them as you evaluate SEUs and stratify them, or as you add submeters to provide additional measures of such items as data centers, heavy machines, or the air-conditioning and heating system.

The total tons of CO_2 from the electricity usage in 2010 was 12,090.672, and for 2012 it was 11,726.2. That represents a reduction of 364.472 tons of CO_2 or 728,802 pounds.

Of course, the official EnPIs are for the year. In actuality, the data are gathered from the monthly electric and gas bills and are placed in a data table underneath a column chart plotted by month.

EnPI: _Electricity reduction + natural gas reduction_

Title: _Energy Performance_

Data source: _Electric and natural gas bills_

Unit of measure: _kBtu_

Frequency of use (circle one):   Monthly   Quarterly   Semiannually   **(Annually)**

Person responsible: _Energy team leader_

Date of data gathering: _As soon as the December electricity and natural gas bills are available_

2010 kBtu – 2012 kBtu = electricity performance

2010: 16,120,900 kWh × 3,412.14 Btu/kWh = 55,006,767,000 Btu = 55,006,767 kBtu

2012: 15,635,029 kWh × 3,412.14 Btu/kWh = 53,348,907,000 Btu = 53,348,907 kBtu

55,006,767 kBtu – 53,348,907 kBtu = 1,657,860 kBtu or **3% reduction** from baseline year

2010 kBtu – 2012 kBtu = natural gas performance

2010: 19,046.25 cu. ft. × 1,020 Btu/cu. ft. = 19,427,175 Btu = 19,427.18 kBtu

2012: 18,105 cu. ft. × 1,020 Btu/cu. ft. = 18,467,100 Btu = 18,467.10 kBtu

19,427.18 kBtu – 18,467.10 kBtu = 960 kBtu or **5% reduction** from baseline year

2012 energy performance is the electricity reduction plus the natural gas reduction

1,657,860 kBtu + 960 kBtu = **1,658,820 kBtu reduction** in energy usage

Figure 4.14   QVS Corporation third EnPI.

> EnPI: _Total energy intensity reduction_
>
> Title: _Energy Performance Intensity_
>
> Data source: _Third EnPI_
>
> Unit of measure: _kBtu/sq. ft._
>
> Frequency of use (circle one):   Monthly   Quarterly   Semiannually   **(Annually)**
>
> Person responsible: _Energy team leader_
>
> Date of data gathering: _As soon as the third EnPI is complete_
>
> 2010 kBtu/sq. ft. − 2012 kBtu/sq. ft. = electricity intensity savings
>
> (55,006,767 kBtu / 480,500 sq. ft.) − (53,348,907 kBtu / 480,500 sq. ft.) = 114.478 kBtu/sq. ft. − 111.02 kBtu/sq. ft. = 3.45 kBtu/sq. ft. or **3% reduction** from baseline year
>
> 2010 kBtu/sq. ft. − 2012 kBtu/sq. ft. = natural gas intensity savings
>
> (19,427.18 kBtu / 480,500 sq. ft.) − (18,467.10 kBtu / 480,500 sq. ft.) = 0.040 kBtu/sq. ft. − 0.038 kBtu/sq. ft. = 0.002 kBtu/sq. ft. or **5% reduction** from baseline year
>
> Total energy intensity reduction = 3.45 kBtu/sq. ft. + 0.002 kBtu/sq. ft. = 3.452 kBtu/sq. ft.
>
> *Note:* This intensity indicator does not change any reduction percentage unless the square footage changed from year 2010 to year 2012, but enables an organization to compare with a similar facility.

Figure 4.15   QVS Corporation fourth EnPI.

ENERGY OBJECTIVES, ENERGY TARGETS, AND ENERGY MANAGEMENT ACTION PLANS

ISO 50001 EnMS Standard

Purpose: O&Ts must be established to support the energy policy, to improve operations and processes, to eliminate energy waste, and to continually improve the organization's energy performance. Every O&T must have an energy action plan that tells who is going to do what, and where.

O&T and energy action plan characteristics: The targets should be SMART: specific, measurable, actionable, reviewable and relevant, and time framed (most O&Ts are from less than one year to two years in duration). The energy action plan should show who is responsible for each task or activity, what is to be achieved when, and when all the activities on the energy action plan have been completed and the O&Ts have been met. Verification as to whether the O&Ts are being completed in an acceptable manner is done by reviews of the responsible person, the energy team, and the management representative and by the management review. All O&Ts and action plans should be documented. A simple template including O&Ts and the energy action plan may be used; in more complex situations, a Gantt chart or critical path method (CPM) chart may be necessary for project management.

Operational Explanation

Establishing O&Ts provides the means and path for putting the energy policy into action. Prior to establishing an objective or target, other preparations have been made. In strategic planning, the voice of the customer has been reviewed and analyzed; the mission statement has been established along with a vision that shows where the company or organization wants to be in 20 years; a strengths, weaknesses, opportunities, and threats (SWOT) analysis has been conducted; values for guiding employees' daily actions have been identified; key result areas have been determined; and goals have been established. Now the Strategic Council is ready to establish objectives. How about the energy team? What does it need to know prior to establishing objectives?

The energy team already knows the company. QVS Corporation has committed both to using ISO 50001 EnMS and to a strategic objective of reducing electricity costs and kWh usage by 10% from the 2010 baseline by 2015. The scope has been established as the five buildings of QVS (the headquarters building, the three plants, and the distribution center) and the outside yards of each. The Strategic Council has committed to performing management reviews at least once a year and executive updates quarterly. It has established an energy policy and communicated it to all QVS employees and contractors. The energy team has been formed, and roles and responsibilities have been established. An electricity and natural gas profile has been established, and all legal and other requirements have been identified. The legal and other requirements should be researched and documented for consideration in establishing O&Ts. The identified prioritized opportunities, SWOT, SEUs, and identified energy efficiencies should be considered in this O&T setting process. The legal and other requirements, technology possibilities, energy-efficiency opportunities, and other suggestions should be considered in developing O&Ts. The targets should be made SMART. At this point, O&Ts are ready to be established that, when completed, will enable QVS to achieve its strategic objective of reducing electricity costs and kWh usage and complying with ISO 50001 EnMS standards. ISO 50001 states that the O&Ts should comply with legal and other requirements and be measurable, significant actions or contributions should be identified in order to meet targets, and responsibilities should be established.

Three Types of Objectives and Developing O&T Statements

There are three types of objectives. The first type is to "improve" something. Improvement can be the maximization or minimization of something, such as "reduces electricity outages." The second type is to "maintain," as in "maintain the transformers that step down the electricity voltage entering our facilities." The third type is to "study/research," as in, "conduct a feasibility study or study the problem to obtain data."

If no target is established, then the objective is made SMART. Objective statements tell what we want to accomplish—a desired outcome, a deliverable, a goal, or an improved product, service, or information. The following are examples of objectives:

- Reduce electricity usage at Plant A by 10% this fiscal year
- Reduce the company's electricity cost by 10% this fiscal year

The target is what you will specifically focus on to achieve the objective. The following are examples of targets:

- Set the thermostat at 78° instead of 72° in the summertime
- Install occupancy sensors in areas of infrequent use
- Implement a "turn off the lights when not in use" policy in all facilities
- Implement an IT power management program on all computers, monitors, and laptops
- Add more insulation in the attic of all QVS facilities
- Use an EnMS to control lights and air handlers in certain locations

Forms, Templates, Processes, and Plans to Meet the Standard

Prior to establishing O&Ts, the energy policy should be developed, the SEUs should be identified, the energy efficiencies should be outlined, and an SWOT analysis may be done. The legal and other requirements need to be developed and considered in the objectives development process. O&Ts can be developed at several different organizational levels, such as the Strategic Council, the energy champion, and the cross-functional energy team. The latter is where most O&Ts should be established. The others will need to approve them eventually and can add to them if needed at that time.

The common objectives process is for the energy team members to review the SWOT analysis and the legal and other requirements first and then list possible objectives. When the list is complete, the team will select those it wants to work on first. If the list is large, it may be helpful to use a criteria matrix to help the selection process (see Table 4.6).

The possible objectives with the highest scores will be selected first. When selected, place a "yes" in the last column. Be careful not to choose too many for the team to develop and implement at one time. After the energy team finishes the current selected objectives, it may later select some of the lower-scored objectives.

Making Targets SMART

SMART means specific, measurable, actionable, relevant, and time framed.

Target 1: Set the thermostat at 78° instead of 72° during the summertime. Is the target specific? Yes. It tells one exactly what to do. Measurable? Yes. One can look at the thermostat to determine whether it is set at 78°. Actionable? Yes. One only has to change the setting. However, prior to this action, it is important to explain to everyone why you are doing this, what you hope to achieve, and why their buy-in is important in achieving the objective. Otherwise you will have a rebellion. Relevant? Absolutely. This change will save a lot of kWh and lower the electricity usage. Time framed? No. There is no indication of when this will be done or how long it will take. To make it time framed, add either a date (e.g., October 1, 2012) or a period of time (e.g., during fiscal year 2013).

Now, the possible objectives need to be placed on a template, with the targets developed first and then the energy action plan for each objective. For convenience, the template includes the objective, the target, and the energy plan on one page (see Figure 4.16).

Table 4.6 Sample form for evaluation and selection of possible objectives.

Possible objective	Supports the energy policy? (1–5 with 1 = somewhat and 5 = highly)	Reduces kWh (electricity) or cu. ft. (natural gas) usage? (1–5 with 1 = somewhat and 5 = significant)	Cost of implementation (1–5 with 1 = high and 5 = low)	Ease of implementation (1–5 with 1 = difficult and 5 = very easy)	Total score	Selected? (yes/no)

Facility name:		Objective #:	
Objective:			
Target:			
Initiation date:	Anticipated completion date:		Actual completion date:
Electricity high users addressed:			
Baseline:		Monitored or measured:	

Energy Action Plan				
Required action	Person responsible	Target date	Status	Comment
#	List each step needed to ensure O&T is met.	Enter a name.	Enter the date the team expects this step to be done.	Enter "red," "yellow," or "green."	Enter the status of this step and record the date beside it (e.g., "Completed [4/4/11]" or "Management has not yet responded, extending target date by 10 days to 4/18/11 [4/4/11]").
1					
2					
3					
4					
5					
6					
7					
8					
9					
10					
11					
12					

Figure 4.16 Objective, target, and action plan template.

Enter the title of the objective and then enter the target. The target is what will be addressed in this objective and, if known, the amount of the target. Then, put in the objective number. Objective numbers start with "OT," followed by the year they were originated and the order in which they were put on a template. For example, the objective number for the first one developed in 2012 would be OT-12-01. Next, the initiation date is the date that the objective was selected. The anticipated completion date is when you expect to have the objective finished. This date can be revised if necessary. It is advisable to revise when necessary so that the auditors or self-inspection does not write it as a problem. If the product or service that results will need to be placed on the monitoring and measuring list later, write "yes" in the "monitored or measured" space. Next, put the baseline for the measure you are using for the objective. If the baseline is not available, then "developing a baseline" should be one of the first actions in the energy action plan.

QVS Corporation Example of Implementation

The facilitator had each team member read as homework the requirements of ISO 50001 EnMS down to establishing O&Ts. The team members discussed the technology advances that should be considered in establishing objectives. They mentioned the improvements in lighting (T5s), new automation equipment for selecting orders in the distribution center, improvements in boilers and HVAC, new building automation or an EnMS for the facilities, IT power management, use of Energy Star and EPEAT equipment and appliances, and large fans to circulate the air, all of which will improve the morale of workers while reducing energy use. The financial condition of the company was discussed with the energy champion. He said the team ought to explore ways to reduce electricity costs with actions that cost little to implement. Then in a few years, add more projects with good returns on investment (ROIs) or a short payback period to further reduce energy use. The financial condition of the company is okay now and is improving each year. Next, the team leader and the facilitator showed what the team had done, including the energy policy, energy profile, and roles and responsibilities established for deploying the strategic objective. The facilitator had the team review the requirements that they were to accomplish in the implementation plan since some can be accomplished by an O&T and some by projects. The facilitator wrote the list on one side of the whiteboard located in the meeting room. The list is shown in Table 4.7.

The facilitator stated that often these requirements are best done by establishing an objective and then adding the target. It is easier if we identify the criteria for turning these into an objective. The criteria are that it must be stand-alone or independent from the other choices, it must be specific and different from the others, and it must lend itself to be an objective. Table 4.8 compares each possible objective with the criteria using a scale of 1–5, with 5 being the highest score. The ones that receive the highest total score will be objectives. The team members assigned the scores.

All objectives should be numbered to facilitate excellent documentation and to make sure that the team and company are using current and updated documents. An easy and clear method of assigning O&T numbers is to start with OT (which

Table 4.7 Implementation requirements being considered as an objective.

Reference and implementation requirement	Criterion 1	Criterion 2	Criterion 3	Total score and selection
Par. 4.5 ISO 50001: Awareness training and competency training				
Par. 4.5 ISO 50001: Design				
Par. 4.5 ISO 50001: Operational controls				
Par. 4.5 ISO 50001: Documentation				
Par. 4.5 ISO 50001: Communication				
Par. 4.5 ISO 50001: Purchasing control plan				

Table 4.8 QVS Corporation implementation requirements being considered as an objective.

Reference and implementation requirement	Criterion 1	Criterion 2	Criterion 3	Total score and selection
Par. 4.5 ISO 50001: Awareness training and competency training	5	5	5	15 (selected)
Par. 4.5 ISO 50001: Design	5	5	2	12
Par. 4.5 ISO 50001: Operational controls	3	4	5	12
Par. 4.5 ISO 50001: Documentation	5	5	3	13 (already set up documentation file, now just do it)
Par. 4.5 ISO 50001: Communication	5	5	5	15 (selected—develop a communication plan)
Par. 4.5 ISO 50001: Purchasing control plan	5	5	3	13 (selected)

stands for *objective* and *target*), add the calendar year, and then add the number of the objective. Therefore, the objectives so far are the following:

- OT-11-01 Develop awareness training
- OT-11-02 Develop competency training
- OT-11-03 Develop a communication plan that covers both internal and external communications
- OT-11-04 Develop and implement a purchase control plan

The team now has four objectives but only one, OT-11-04, that may reduce electricity costs and usage. The facilitator asked the team, "Do any of you have any ideas of how we can reduce electricity costs and kWh usage?" John said if we could get everyone to turn off the lights when they leave a room, we could reduce electricity usage. Larry stated that we need to implement power management on our computers, monitors, and laptops. The team leader said, "Excellent ideas. We need to turn your two ideas into objectives to develop and implement an electricity conservation program and an IT power management program." The team agreed and added "OT-11-05 Develop and implement an electricity conservation program" as the fifth objective and "OT-11-06 Implement an IT power management program" as the sixth objective. The team leader stated that the team needs to do two other actions. The first is to determine the major users of electricity, a requirement of ISO 50001 EnMS. The second is to identify projects that, when implemented, will reduce kWh usage and cost. He stated that an expert should perform an energy audit of the four facilities and recommend projects or actions for reducing electricity usage. For each project or action, the expert should determine the payback period so that the team can prioritize them to get the most "bang for its buck." The team leader will talk with the energy champion on this to gain his approval and support because funds will be needed. If he approves, then the energy audit will be OT-11-07. The next meeting will focus on identifying the major energy users at the facilities. John and Ted volunteered to be the responsible individuals for OT-11-01. Mary from accounting volunteered to be the responsible person for developing the purchase control plan (OT-11-04). Ann volunteered to be the responsible person for the communications plan (OT-11-03). Bill stated that he and the points of contact (POCs) at three of the facilities (who are facility managers) would develop the energy conservation plan (OT-11-05) and also manage the energy audits (OT-11-07). The other two facility managers who are not on the team will be brought in on the energy audits for their facilities. The team leader said that at the next meeting, all team members should be ready to report on what they have accomplished and should develop an action plan for their specific objective. The objectives got the energy team moving toward the implementation stage. Each O&T must be shown on a template like Figure 4.17 and then the energy action plan needs to be developed.

Does the O&T shown in Figure 4.17 meet the SMART criteria? The target is QVS Corporation's employees and contractors. Is that specific? Yes, it is. Measurable? Yes. Either the awareness training is done or it is not. Also, was the kWh consumption reduced once everyone had the training? Is it actionable? Absolutely. Develop and present the training. Relevant? Yes. It fits the EnMS well in that it

Facility name: QVS Corporation, Gun Barrel City, Texas			Objective #: OT-11-001	
Objective: To develop a an electricity conservation program and kWh awareness training program.				
Target: Employees and contractors at QVS Corporation				
Initiation date: 1/2/2011		Anticipated completion date: 12/31/2015		Actual completion date:
Electricity high users addressed: AC and heating, lighting, office equipment including computers and monitors, and other				
Baseline: 2011 electricity costs and kWh used by all facilities			Monitored or measured:	

Energy Action Plan

#	Required action List each step needed to ensure O&T is met.	Person responsible Enter a name.	Target date Enter the date the team expects this step to be done.	Status Enter "red," "yellow," or "green."	Comment Enter the status of this step and record the date beside it (e.g., "Completed [4/4/11]" or "Management has not yet responded, extending target date by 10 days to 4/18/11 [4/4/11]").
1	Identify actions that can be taken to save energy.	EnMS team	2/3/2011	Green	
2	Select actions for electricity conservation program.	EnMS team	2/16/2011	Green	
3	Get actions approved by objective champion.	Team leader	2/20/2011	Green	Objective champion briefed Strategic Council and it liked the program.
4	Put actions into a PowerPoint presentation.	Bill Gibson	2/22/2011	Green	
5	Send to all employees and contractors.	Objective champion	3/2/2011	Green	

Figure 4.17 QVS Corporation O&T action plan.

contributes to QVS Corporation reaching its energy goals. Time framed? Yes. It needs to be done by December 31, 2015.

Each O&T should be checked for SMART as a template is developed for each objective. If the target is not SMART, figure out how to make it SMART by meeting each criterion.

ENERGY ACTION PLANS AND PROJECTS

With some objectives identified, the facilitator knew that action plans were necessary to identify who is going to do what, where, and when. The facilitator had a training session for developing action plans. The energy team and the POCs at the five facilities attended. The facilitator first introduced the action plan template.

The team filled in the action plan. It was reviewed and the status was assigned ("green" in each step) later as the actions happened and were completed. A Microsoft PowerPoint presentation for an electricity conservation program was developed.

Of course, these O&Ts just got the EnMS going. Other O&Ts were established as the program was developed. However, an O&T and action plan template should still be completed for each O&T as it is originated. OT-11-04 Buy Energy Star and EPEAT, OT-11-05 Implement an IT power management program for computers, monitors, and laptops, OT-12-01 Reduce energy used for air conditioning the data center, OT-12-02 Reduce cooling energy use in all facilities, and OT-12-03 Reduce energy used to heat all facilities. For the first goal of reducing electricity usage by 10% by 2015, the energy conservation plan, when implemented, should reduce electricity usage by 5% and IT power management by 5%, giving the team sufficient contribution to meet the goal or target. In the future, projects will need to be identified, approved, advertised, and implemented that have a good rate of return (savings per kWh reduction in a year). Either doing an energy audit or prioritizing the energy efficiencies by facility is the best option for identifying the projects. After identifying the projects, estimate how much each project will cost and how much contribution (kWh or cubic feet) it will reduce, and then get the payback in years (see Figure 4.18). Select projects with ideal payback periods first (three years or less), and over time, executive orders state it is feasible to fund projects with a payback period of up to 11 years.

Projects that are contracted need a statement of work and a schedule for completion, which is also an action plan. Therefore, both O&Ts and contracted projects are called projects by the teams. A contracted project could also have an O&T and an action plan. The projects identified for possible implementation to further reduce energy use are shown in Table 4.9.

QVS Corporation would first select projects with payback periods of less than three years and ensure that they add up to the contribution amount needed to reach the goal or target. If they do not, then a project should be selected from a higher year's payback that will have a sufficient contribution. If further reduction is desired, then a project should be selected from the next level: projects with payback over four years. Projects like those shown in Table 4.9 can come from several sources, but conducting an energy audit every three years is probably the best option.

Now, it is time to start implementing.

Example: Hangar light change-out (from metal halite lights to T5 lights)

Current: 120 metal halite lights are on 9 hours/day for 5 days/week and each is 1,000 watts
Operating time = 52 weeks/year × 5 days/week × 9 hours/day = 2,340 hours/year
Energy usage = 120 lights × 1,000 watts/light = 120 kW × 9 hours/day =
1,080 kWh/day
Cost = 1,080 kWh/day × $0.0679/kWh = $73.33/day

120 lights × 1,000 watts = 120 kW
120 kW × 2,340 hours/year = 280,800 kWh/year

Future: 120 T5 lights are on 9 hours/day for 5 days/week and each is 450 watts
Operating time = 52 weeks/year × 5 days/week × 9 hours/day = 2,340 hours/year
Energy usage = 120 lights × 450 watts/light = 54 kW × 9 hours/day = 486 kWh/day
Cost = 486 kWh/day × $0.0679/kWh = $33.00/day

120 lights × 450 watts = 54 kW
54 kW × 2,340 hours/year = 126,360 kWh/year

Present − future = 280,800 kWh/year − 126,360 kWh/year = 154,440 kWh/year
154,440 kWh/year × $0.0679/kWh = **$10,486.48 savings**

Figure 4.18 QVS Corporation sample payback estimate.

Table 4.9 QVS Corporation funded projects.

Facility	Project	Cost	Payback period	kWh reduction per year (%)	Remarks
HQ	Balance air flow with new air handlers and building automation system (BAS)	$40,000	3 years	3	
	Turn off air to facilities not in use after normal hours	$0	Immediately	1	
	Replace boiler	$120,000	10 years	1.5	
	Replace T12 lights with T5 lights	$103,000	7.8 years	14	
	Implement IT power management	$4,000	1 year	6	Computers, monitors, laptops
	Install occupancy controls (OCs)	$6,810	3.2 years	0.5	Any place not used frequently

(continued)

Table 4.9 QVS Corporation funded projects. *(Continued)*

Facility	Project	Cost	Payback period	kWh reduction per year (%)	Remarks
HQ (cont.)	Separate hot and cold aisles in data center to reduce A/C furnished	$38,000	2 years	10.5	Increase PUE to under 2
Plant A	Replace T12s lights with LEDs	$150,000	8.2 years	12	
	Place additional insulation in roof attic	$45,000	3 years	6	
	Add advanced electricity meters to the BAS	$16,000	3 years	5	
	Add capacitors to reduce power factor fees	$16,200	3 years	0	Eliminates power factor adjustment cost but does not save any kWh
	Add OCs to break rooms and mechanical and electrical rooms	$7,000	3.3 years	0.5	
Plant B	Replace T12s lights with LEDs	$140,000	7.6 years	11	After installation is completed
	Place additional insulation in roof attic	$37,000	3 years	5	
	Add advanced electricity meters to the BAS	$6,200	3 years	6	
	Add capacitors to reduce power factor fees	$15,500	3 years	0	Eliminates power factor adjustment cost but does not save any kWh
	Add OCs to break rooms and mechanical and electrical rooms	$6,000	3.3 years	0.6	
	Replace doors leaking air	$2,100	1 year	0.3	
	Repair cooling tower	$28,000	5 years	2	
Plant C	Replace T12s lights with LEDs	$140,000	7.6 years	1	After installation is completed

Table 4.9 QVS Corporation funded projects. (Continued)

Facility	Project	Cost	Payback period	kWh reduction per year (%)	Remarks
Plant C (cont.)	Place additional insulation in roof attic	$6,200	3 years	6	
	Add advanced electricity meters to the BAS	$16,000	3 years	5	
	Add capacitors to reduce power factor fees	$15,500	3 years	0	Eliminates power factor adjustment cost but does not save any kWh
	Add OCs to break rooms and mechanical and electrical rooms	$6,000	3.3 years	0.6	
	Replace doors leaking air	2,100	1 year	0.3	
	Add big fans in main production and warehouse area	$52,000	3 years	1.2	Will reduce natural gas to the heaters by 33% a year
Plant D	Replace T12s lights with LEDs	$140,000	7.6 years	11	After installation is completed
	Place additional insulation in roof attic	$37,000	3 years	5	
	Add advanced electricity meters to the BAS	$16,000	3 years	5	
	Add capacitors to reduce power factor fees	$15,500	3 years	0	Eliminates power factor adjustment cost but does not save any kWh
	Add OCs to break rooms and mechanical and electrical rooms	$6,000	3.3 years	0.6	
	Replace doors leaking air	$2,100	1 year	0.3	
	Replace exit signs	$1,200	2.5 years	0.03	
	Upgrade security lighting	$8,100	7 years	1.6	

Chapter 5
Implementation and Operations

AN OVERVIEW
ISO 50001 EnMS Standard

Purpose: The standard calls for plans or processes that may come out of the planning phase, but if they were not developed then, they must be developed now.

Implementation characteristics: Training plans that cover both awareness and competence must be developed and accomplished. A communication plan that includes guidance on internal and external communications should be developed and deployed. A document control and control of records plan should be developed and communicated, and a document control system should be put into place, such as an organization's management system or an intranet system. Operational controls should be developed for each SEU. Energy consideration should be a requirement in the design of processes, systems, equipment, and facilities. Plans or policy should require purchase and use of energy-efficient equipment and products.

Operational Explanation

Now that the planning is done, the energy champion and the energy team are ready to implement the EnMS throughout the organization. A few standard requirements must take place in the implementation phase if they have not been done previously: having a procurement plan that includes procurement buying energy-efficient electronics and equipment, having a communications plan that provides guidance on communicating internally and externally, and providing awareness training to all employees and contractors of the facility. Also, for management to fully support implementation of the EnMS, it must be aware of the standard and what it entails. Personnel included in the EnMS implementation and later maintenance must be competent and qualified to do the tasks required of them. The document control and control of records plan, to include identifying nonconformances and corrective action, must be in place. A monitoring and measurement plan should have been developed in the planning phase, and now it should be in place and monitored.

Forms, Templates, Processes, and Plans Required to Meet the Standard

All the forms developed in the energy planning and review phase or processes are used in this chapter to implement the EnMS.

QVS Corporation Example of Implementation

QVS Corporation used the EnPIs and the 2010 energy consumption baseline to establish energy goals. Energy efficiencies were identified and used to establish O&Ts and projects. SEUs were identified and monitored.

COMPETENCE, TRAINING, AND AWARENESS
ISO 50001 EnMS Standard

Purpose: To ensure that people working in energy-related processes or jobs are competent to perform their tasks and that all employees are aware of the EnMS.

Competence, training, and awareness characteristics: EnMS awareness training should be provided to all management, employees, and contractors. The training should include the energy policy, procedures, O&Ts, the benefits of improved energy performance and how each individual can contribute, the roles and responsibilities of the persons responsible for achieving the requirements of the EnMS, and other pertinent data. Training can be provided in the classroom, on the job, or in other forms to ensure competence in such areas as analyzing energy bills, reading meters, learning the equipment, identifying the SEUs, and reading current and voltage, power factor, load factor, and other useful indicators. All training should be documented in the organization's training records.

Operational Explanation

All employees and contractors of the organization should be trained in the energy roles and responsibilities associated with their jobs. The best way to accomplish this is to develop an energy awareness training package (using a Microsoft PowerPoint presentation is recommended). It is a best practice to include energy conservation actions that all individuals can take to reduce energy consumption, such as turning off the lights when they are no longer needed. At a minimum, the training should include the following:

- The deployment organization, including the Strategic Council, the energy champion, the energy team, and the facility managers, along with their roles and responsibilities
- The five phases of ISO 50001 EnMS
- The strategic goal established, if one has been
- The organization's energy policy

- The EnPIs
- The SEUs and the processes for which employees have direct responsibility that impact energy performance and how important it is for them to accomplish (energy conservation actions and processes)
- The current O&Ts, their impact on energy performance, and their status
- Methods for everyone to follow to reduce energy use

The energy champion and the energy team should undergo this training first and make any needed improvements before it goes to everyone. The energy team leader and the energy champion should make the presentation to the Strategic Council or top management so that they are familiar with what is going on and understand why their support and involvement are crucial to the system's success. It is absolutely essential that an experienced facilitator with ISO 50001 EnMS knowledge be part of the energy team so that each meeting is facilitated.

If any individual needs training to perform his or her role, competency training must be provided. The organization is required to identify the training needs and competency of the staff. In order to demonstrate this, the energy team should develop a training matrix and get the energy champion to approve it. Training records for QVS Corporation employees are kept in human resources in Room 210 of the headquarters building, 2600 Mulberry St., Gun Barrel City, Texas. Although groups are used in the training matrix, the individuals in those groups will have their training records annotated with the training they receive.

Once the energy conservation techniques have been presented and everyone has received awareness training, additional training should be provided to the facility managers at the organization. As the energy managers at the individual facilities, they can benefit from attending an energy management conference or seminar or participating in webinars. Having the energy team leader and the facilitator develop and present additional training would also be beneficial to improving energy performance. The course should delve deeper into the billing statements, including the different adjustment fees and charges, so that the facility managers can analyze the statements and be sure that the organization is billed correctly. The major formulas used in measurement are used to develop graphs that show progress toward the target. The SEUs and how the energy consumption has been reduced at other organizations are benchmarked and improvements are made in the organization's methods and processes.

Forms, Template, Process, and Plans to Meet the Standard

Table 5.1 shows an example of a training matrix/plan.

QVS Corporation Example of Implementation

The energy team developed a training plan that included both competency and awareness training (see Table 5.2).

Table 5.1 Sample organization's training plan/matrix.

Name or group	Role(s)	Required training	Competent? (yes/no)	Training received to date	Date when refresher training is deeded	Remarks
Strategic Council/ top management	Leadership, guidance, approval					
Energy champion	Leadership, technical advisor, approves plan and presents program to management and employees					
Energy team	Plan, develop, implement, maintain, and sustain EnMS					
Facilities managers and staff	Energy manager for their facility					
Electricians and mechanical technicians	Read submeters and spot meters on SEUs, implement maintenance and minor construction projects concerning SEUs					
Administrative staff	Use power management, use duplex printing, turn off lights when not in use					
Plant employees	Produce quality products using less energy					

Table 5.2 QVS Corporation training plan/matrix.

Name or group	Role(s)	Required training	Competent? (yes/no)	Training received to date	Date when refresher training is deeded	Remarks
Strategic Council/ top management	Leadership, guidance, approval	EnMS awareness Energy awareness	Yes	EnMS awareness (1/5/2011) Energy awareness (3/10/2011)	January 2013 March 2012	See attachment #8 for PowerPoint EnMS awareness training
Energy champion	Leadership, technical advisor, approves plan and presents program to management and employees	EnMS awareness, energy awareness	Yes	EnMS awareness (1/5/2011) Energy awareness (3/10/2011)	March 2012	See attachment #9 for energy awareness training
Energy team	Plan, develop, implement, maintain, and sustain EnMS	EnMS awareness, energy awareness	Yes	EnMS awareness (1/5/2011) Energy awareness (3/10/2011)	March 2012	Ongoing research and benchmarking adds to their education
Facilities managers and staff	Energy manager for their facility	Energy awareness, energy manager training	Yes	Energy awareness (3/10/2011) Energy manager training (7/15/2011)	March 2012	Individual processes involving SEUs are discussed with applicable plant personnel
Electricians and mechanical technicians	Read submeters and spot meters on SEUs, implement maintenance and minor construction projects concerning SEUs	Energy awareness, meter vendors on-the-job training (OJT)	Yes	Energy awareness (3/10/2011) OJT with meter vendors (2/11/2011)	March 2012	
Administrative staff	Use power management, use duplex printing, turn off lights when not in use	Energy awareness	Yes	Energy awareness (3/10/2011)	March 2012	
Plant employees	Produce quality products using less energy	Energy awareness	Yes	Energy awareness (3/10/2011)	March 2012	

COMMUNICATIONS

ISO 50001 EnMS Standard

Purpose: To communicate EnMS happenings to all management, employees, and contractors.

Communications characteristics: Establish a process so that all can contribute suggestions for improvement, and have a plan for handling both internal and external communications.

Operational Explanation

ISO 50001 EnMS standard requires that a process be developed and implemented to manage internal and external communications to ensure that employees are committed to the energy policy and that they accept and participate in energy O&Ts, and to ensure that accurate information is given for any external request for data and information. To meet the standard, the organization needs to know who is responsible for communicating information about the EnMS and the communications method and/or media that are to be used. Possible communication methods are:

- E-mails and bulletin boards
- Organization videos and intranet communications
- Staff publications such as magazines or papers
- Staff meetings, town hall meetings, seminars, and tool box meetings and discussions
- Awareness days, campaigns, and kaizen events aimed at improving a SEU

The type of information can be the energy policy, the energy team, O&Ts, how employees can help, power management, success stories, monthly electricity and natural gas consumption, whether the organization is meeting its goals, dollar savings, and contact points for employees to offer suggestions or voice complaints. All employees and contractors working at the facility should be encouraged to contribute suggestions and to make comments about what they see that has changed or improved and about the EnMS and energy performance.

Management decides whether communications will take place outside the organization. QVS Corporation's top management has said, "Yes, we are proud of what we are doing and maybe our sharing of the information will encourage others to do so. If we do it right, then this is an excellent opportunity to improve our public image." The communication plan should show who is responsible for communicating information on the EnMS and energy performance, what information can be communicated, the means of communicating, who will keep the records, and where the records will be kept.

Forms, Template, Process, and Plans to Meet the Standard

The most important area here is to have a plan written in the format of the organization's other plans. It should show the purpose, processes of communicating, and who needs to approve what type of communications so one knows and can adhere to the communications plan. The plan format should contain at least these major areas:

- Purpose
- Requirement/scope
- Definitions
- Responsibilities
- Internal communications process
- External communications process
- Points of contact

QVS Corporation Example of Implementation

Figure 5.1 shows QVS Corporation's communications plan.

Memorandum

Subject: QVS Corp. Energy Communications Plan   Date: January 27, 2011
To: All QVS Corp. personnel
From: Gene Smith, President, QVS Corp.

Purpose
This memorandum describes QVS Corporation's policy for both internal and external communications in regard to QVS Corp.'s Energy Management System (EnMS). The specific purpose is to spell out the methods and processes for communicating with QVS Corp.'s management, employees, and contractors, and instruct how to handle external inquiries for information.

Requirement/Scope
This memorandum is in response to paragraph 4.5.3 Communications, ISO 50001 EnMS.

Definitions
Communications media: A medium is one specific method of communicating. E-mail is a medium. A letter is a medium. An announcement in a staff meeting is a medium. Other media include signage, pamphlets, handouts, in-person training, loudspeaker announcements, intranet, notice boards, bulletin boards, kaizen events, town hall meetings, seminars, training sessions, and telephone calls.

EnMS: The Energy Management System in accordance with ISO 50001 Energy Management System (EnMS).

Major energy concerns: Concerns that deal with legal requirements not being implemented or followed that could lead to adverse situations, for example, the potential loss of power for prolonged periods due to weather or legal concerns or a contingency plan becoming inoperable for some unforeseen conditions. Major energy concerns can be considered emergency communications.

Figure 5.1   QVS Corporation's communication plan.   *(continued)*

Minor energy concerns: Concerns that deal with energy issues that can be addressed or improved, but have no imminent health, environmental, loss of energy, or legal repercussions. Minor energy concerns can be dealt with through a facility's or organization's energy team as necessary. Examples are any energy information that does not require immediate action to solve or disseminate. The answer can be obtained during the next three days and the issue does not in any way become an emergency situation or problem.

Formal energy requirement: A letter or e-mail request for information in regard to a procedure, audit, or similar request that spells out what is desired and by when.

Informal energy requirement: A telephone or in-person request for specific information that may or may not have an associated timetable. The request could come from a QVS Corp. person or an external person.

POC: Primary point of contact.

Staff: The people who work at the facility, including management and employees.

Responsibilities

The energy champion or his/her designee is responsible for documenting and keeping records of internal and external major energy concerns that reach his/her level for action or forwarding to the QVS Corp. Strategic Council. The energy champion will be informed of all major energy concerns and all minor concerns.

The EnMS or energy team is responsible for keeping QVS Corp.'s personnel and contractors informed on EnMS-related matters, to include how they can help improve its energy performance. The EnMS team will keep records of all internal or external, major or minor energy concern communications that they deal with, including any responses. The documentation will be in the share drive EnMS file in the communications folder under internal or external communications as appropriate. The files on the share drives are available to the staff for information as read-only. These files must be kept current by the EnMS document control manager. Letters, standard operating procedures (SOPs), work instructions, and company procedures are other communication tools available.

The energy champion is responsible for providing a quarterly update to the Strategic Council on EnMS progress, barriers, and performance. The energy team is responsible for assisting the champion in preparing the quarterly updates.

The corporate energy team is responsible for providing annual energy awareness training and any competency training needed to ensure process and objective performance.

The building's facility manager is responsible for handling all energy matters and answering energy questions for for his/her facility. In the event of a major energy concern, the facility manager will notify the corporate energy team leader immediately.

Internal Communications Processes

Sending communications: The EnMS or energy team will communicate with the staff using any appropriate media. To ensure that the message reaches everyone, using more than one medium is often required.

Receiving communications:

Minor energy concerns: Any minor energy concerns, suggestions, ideas, problems or perceived problems, or other EnMS-related communications should be reported to the energy team leader or any EnMS team member. Concerns relating to occupational health and safety aspects should be given to the safety officer, who will either address them or elevate them to the level where they can be solved.

Major energy concerns: Major energy concerns should be immediately directed to a supervisor, manager, vice president, or a Strategic Council member. As appropriate, they will respond to

Figure 5.1 QVS Corporation's communication plan. *(Continued)*

communications. Any concern that cannot be addressed at the facility level will be directed to the corporate energy team leader or to the energy reduction objective champion.

External Communications Processes
Receiving communications: Any QVS Corp. employee could receive an external request for information by receiving an e-mail, a letter, or a telephone call or by having a visitor. This request should be made known immediately to a supervisor or corporate energy team leader, if energy related, who in turn will bring to the energy reduction objective champion's attention (all minor or major energy concern external requests). The energy champion will respond or send to the appropriate person or office for response.

The energy champion will determine which media inquiries should be referred to the Freedom of Information and Records Management Section at QVS Corp.'s headquarters.

All external inquiries with energy issues by members of the media or the public must be forwarded to the appropriate public information officer through the chain of command. Legal communications concerning the energy performance of the facility will be handled by the energy champion.

Points of Contact
The energy team leader is always a point of contact for minor energy concerns. In the event of a major energy concern, the employees should immediately report to their supervisor, who in turn will resolve or present to the energy champion for resolution.

Communication is the key to achieving our mission and meeting our EnMS responsibilities. Your past communications, job achievement, and conscientious attitude are most appreciated and needed in achieving our future responsibilities and mission. Energy is the responsibility of everyone.

Figure 5.1 QVS Corporation's communication plan. *(Continued)*

DOCUMENTATION

ISO 50001 EnMS Standard

Purpose: The organization must maintain a central document control system with files covering all 23 elements of the EnMS and other areas deemed important for managing and auditing the EnMS.

Documentation characteristics: At a minimum, the documentation should include the boundaries and scope of the EnMS; the organization's energy policy; all energy O&Ts and energy action plans; meeting minutes, agendas, and sign-in sheets; audits and self-inspections; and other documents required by the ISO standard, such as management reviews' inputs and outputs, evaluations of legal compliance, all SEUs, variables, EnPIs, and other key information.

Operational Explanation

All documents in the organization's EnMS should be developed to ensure they are easy to retrieve, easy to read, and current, meaning revisions should be annotated. The documents or records need to be either numbered or at least dated with the subject so they can be easily retrieved. The documentation and document control plan is shown later in Figure 5.2.

An EnMS manual that includes the organization's procedures and plans is advisable, allowing quick access to the required documents. Instead of the different plans outlined in this chapter and others, procedures would be written and coordinated, and approvals obtained and listed in this manual.

Forms, Template, Process, and Plans to Meet the Standard

A central document control system needs to be established on the organization's management operating system or on Microsoft SharePoint. File categories should be established that guide where something should be filed and provide structure to the documentation process. The following file categories are recommended:

- EnMS scope and boundaries
- Management representative (include the letter appointing the management representative, the energy team charter, and updates given to the Strategic Council)
- Energy policy
- Legal and other requirements
- Energy team meetings (agendas, minutes, sign-in sheets, and current team profile sheet)
- Significant uses of energy (include SEUs, variables, any correlations, operational controls, etc.)
- EnPIs (include updates of graphs, baselines, and energy profiles of each facility)
- O&Ts and energy plans by year of origination
- Projects (project management records, payback period, energy conservation measures [ECMs], and measurement and verification)
- Communications plan
- Document control plan
- Design (show how energy reduction items were used in design)
- Procurement of equipment and electronics plan (include power management before and after)
- Monitoring and measurement plan
- Legal and other requirements evaluations (show each evaluation by year and any changes or corrections made)
- Self-inspections/internal audits (include completed self-inspection checklist and any CARs or PARs and show resolutions)
- Training (include the training matrix, Microsoft PowerPoint or any other training given, sign-in sheets for the training, and future training under consideration)
- Management reviews (include, by year, the agenda, Microsoft PowerPoint presentations, the minutes, and the sign-in sheets)

QVS Corporation Example of Implementation

Figure 5.2 shows QVS Corporation's documentation plan.

> **QVS Corporation Document and Record Filing System**
>
> The documents and records will be filed on share drive for easy access by the energy team members, the objective champion, and members of the Strategic Council. Only the energy team leader, the energy team document control officer, and the lead facilitator will have the right to file and change anything in the system. The others will have read-only access.
>
> The files will be:
>
> 1. Energy Policy
> 2. Meetings—Minutes, Agendas, and Sign-in Sheets
> 3. Significant Energy Users
> 4. Energy Profile
> 5. Energy Performance Indicators
> 6. Objectives and Targets
> 7. Projects
> 8. Management Reviews
> 9. Self-Inspections
> 10. Legal Requirements
> 11. Nonconformities and Corrective Actions
> 12. Objective Champion Updates
> 13. Communications—Internal and External
> 14. Miscellaneous
>
> O&Ts and projects will be numbered. Meeting minutes and communications will be filed by date and by year originated.
>
> The document control officer or manager controls the access and changes the Microsoft SharePoint program as needed. He or she, along with the energy team leader and the lead facilitator, will ensure that the documents used by the team are current, legible, and traceable. The files will be made available to auditors conducting a second-party audit. The records in the files are to be maintained for four years, and then they should be deleted or archived.

Figure 5.2 QVS Corporation's documentation plan.

CONTROL OF RECORDS

ISO 50001 EnMS Standard

Purpose: All documents required by the ISO standard must be controlled to ensure that only current information is used and that obsolete documents are retired when possible.

Control of records characteristics: All documents must be approved prior to release, reviewed periodically, and updated (marked with new dates and a revision number). The organization must ensure that documents are current, relevant, and available for use (in a centralized document control system, with potential users having, at a minimum, read-only access); all documents are legible and readable; documents essential to the EnMS are controlled; and obsolete and other documents are retired in accordance with retention and retirement criteria and policy.

Operational Explanation

Records that are no longer in use are still needed for documentation. The document control manager must ensure that changes, revisions, and updates are included in the document control system. All documents should be legible and readable. Once a document is over four years old (the retention period), it should be treated as obsolete and disposed of in accordance with the organization's retention and disposal plan.

Forms, Templates, Processes, and Plans to Meet the Standard

Records are filed in the central documentation system on Microsoft SharePoint and are retired at the designated end of their retention.

QVS Corporation Example of Implementation

Figure 5.2 addresses control of records and is an example of QVS Corporation's implementation.

OPERATIONAL CONTROLS

ISO 50001 EnMS Standard

Purpose: To establish operational controls for SEUs and develop an emergency plan for energy.

Operational controls characteristics: Establish operation and maintenance criteria for SEUs and for the way facilities, processes, systems, and equipment are operated and maintained, and communicate these criteria to all involved in this area.

Operational Explanation

For every SEU, an operational control or O&T must be developed. SEUs should be operated and maintained properly. In order to ensure this, criteria such as the following need to be established: writing how systems and equipment should be operated and identifying those settings and running conditions that could affect energy performance and maintenance. The organization must be able to keep some essential operations running so that if there is an electrical power outage, the lights and the air-conditioning or heating in these areas continue. Data centers require redundancy so that their servers don't turn off or the cool temperature needed to properly operate them is not lost. It is not uncommon for an organization to have a separate feeder to the data center and backup generators or uninterruptible power supplies (UPSs) for the data center and other applications. The mission must always go on. An evaluation of what is needed to accomplish this, what should be placed in the plan, and how the plan should be executed should be a necessity for all companies. QVS Corporation's contingency plan is shown later in Figure 5.3. The operational controls are shown first, followed by the contingency plan.

Forms, Templates, Processes, and Plans to Meet the Standard

There are several methods that can be used in documenting operational controls. They could be placed in a maintenance program or procedure. The most common display is to show each SEU with the operational controls that apply. This allows a quick assessment and guide for training of personnel associated with the SEUs. A contingency plan is needed to show what to do in case electricity is lost due to a natural or unnatural situation or disaster.

QVS Corporation Example of Implementation

Tables 5.3 and 5.4 and Figure 5.3 show examples of QVS Corporation's operational controls.

Table 5.3 QVS Corporation's SEU operational controls—electricity.

Major energy users	SEUs	Operational controls
HVAC		
Heating	Boiler	Treating makeup water to prevent equipment damage and efficiency losses. Check the air to fuel ratio to ensure that the combustion process is operating efficiently. Use ASHRAE standards for operating.
Air-conditioning	Air conditioner	Regularly check hoses and valves for leaks, and make any repairs if needed. Cleaning intake vents, air filters, and heat exchangers regularly will increase both equipment life and efficiency. Change filters when needed. Condenser coils should be checked for debris and cleaned on a quarterly basis.
Air-conditioning	Chillers	Use ASHRAE standards for operation. Periodically check sensors that automatically turn unit on and off. Perform preventive maintenance on chiller as recommended by the manufacturer.
Air-conditioning—data center	CRACs	Monitor temperature and humidity gauges.
Air-conditioning	Economizers	Check linkage on damper to see if it is working (sticks and breaks easily).
Ventilation	Roof ventilators	Ensure that ventilators are secured to the roof surface with screws. Inspect exhaust fans to ensure that they run only when needed. Perform routine maintenance.

(continued)

Table 5.3 QVS Corporation's SEU operational controls—electricity. *(Continued)*

Major energy users	SEUs	Operational controls
Lighting	Office lights	Relamping program. Replace old bulbs with energy-friendly bulbs. Replace old ballasts with new electronic ones.
	Plant lights	Relamping program. Replace old bulbs with energy-friendly bulbs. Replace old ballasts with new electronic ones.
	Security lights	Relamping program. Replace old bulbs with energy-friendly bulbs. Replace old ballasts with new electronic ones.
	Exit lights	Replace with more energy-saving models. Replace when not lit.
Office machines	Computers, monitors, and laptops	Use software to check on power management settings. Include specifications to buy energy-efficient electronics.
	Imaging machines (copiers, printers, and fax machines)	Use software to check on power management settings. Include specifications to buy energy-efficient electronics.
	TVs and other electronics	Include specifications to buy energy-efficient electronics.
Motors, machines, shop equipment, automated distribution center		Check for voltage imbalance. Check for cleanliness. Lubricate when needed. Check to ensure it is not running when it should be off.

Table 5.4 QVS Corporation's SEU operational controls—natural gas.

Major energy users	SEUs	Operational controls
Process heat	Furnace	Post signs to keep doors closed. Monitor temperature settings.
Heat	Overhead heaters	Monitor temperature on the floor. Monitor heating setting on control thermostat.
Hot water	Hot water heaters	Monitor temperature settings on hot water heaters. Monitor temperature of water that has been heated.

> *Requirements*
>
> ISO 50001 EnMS states, "The organization shall establish, document and maintain a procedure for identifying and responding to any energy supply or other potential disasters.
>
> This procedure shall seek to prevent or mitigate the consequences of any such occurrence and consider the continuity of the business operations."
>
> *Specific Actions Taken to Mitigate Loss of Power*
>
> In 2010 QVS Corp. experienced numerous service interruptions due to weather and to trees and squirrels interfering with power lines. Also, the power quality was not to the company's satisfaction since it negatively affected the motors and machines in the plants. QVS Corp. convinced the Gun Barrel City Electric Company to run a designated feeder (QVS Corp. was the only company on the feeder) from its new substation to the QVS Corp. complex. The old feeder remained. It was hooked up to another substation fed by another power plant. This arrangement significantly reduced the probability of QVS Corp. facilities losing power.

Figure 5.3 QVS Corporation's energy contingency plan.

DESIGN

ISO 50001 EnMS Standard

Purpose: The organization shall consider the design of new, modified, and renovated facilities, equipment, systems, and processes that can have a major impact on energy consumption, their potential energy performance and use, and tools, equipment, and techniques to achieve energy efficiencies.

Design characteristics: Design improvements shall be incorporated into the work specifications and procurement actions and documented in the document control area under "design."

Operational Explanation

Energy performance should be considered in new or rehabilitated facilities and equipment in the design stage. If developing a construction partnering agreement, it should be one of the goals agreed on by the partnering team including the owner, the contractor, the architect engineer, and all the subcontractors. Using LEED (leadership in energy and environmental design) criteria is a best practice in the design of new buildings. A conceptual model of energy performance is shown below. The analysis of SEUs and the use of energy efficiencies in the design can help the new facility perform better.

ECMs are to be tracked and visible for all to help prioritize funding, schedule appropriate ones for design, and keep track of possible energy projects. The ECM form is shown later in Figure 5.4, followed by a completed ECM (see Figure 5.5).

LEED was developed by the US Green Building Council as a tool to drive construction of green facilities. From individual buildings and homes to the many facilities of an organization, LEED is transforming the way owners think about how their buildings are designed, constructed, maintained, and operated across the globe. LEED is a green building tool/technique that addresses the entire building life cycle and recognizes best-in-class building strategies.

At its core, LEED is a program that provides third-party verification of green buildings. Building projects satisfy prerequisites and earn points to achieve different levels of certification. LEED is driving building construction to develop facilities that furnish enough electricity themselves to meet the needs of the facility's occupant.

Forms, Templates, Processes, and Plans to Meet the Standard

A project plan should be developed by the energy team and the facilities during the planning phase to show major event milestones or completed phases, the date, by whom, where, and what.

Government and other organizations develop ECMs for potential energy projects and track these measures until they become a project. An ECM typical of the ones used in both government and private businesses is shown in Figure 5.4.

The payback period tells how long it will take once the project is completed to recapture the cost of doing the project. The estimated savings is the most difficult part of the form. It satisfies the critical success factor (CSF) "Have sufficient

Energy Conservation Measure		
ECM number:		
Facility and location:	Project title:	Date:
Project originator (name and telephone):	Estimated savings:	Payback period:
Project description/cost:		
Estimated savings calculations:		
Payback period calculations:		

Figure 5.4 Sample ECM.

contribution to meet the target." For each project, an estimate of how many kWh will be saved needs to be calculated. In some industries, past estimates or results will be helpful. A conservation program will save around 5% if implemented. A power management program that is implemented for computers and monitors can save up to $75 a year in electricity for each unit. Separating the hot air from the cold air in the flow of air back to the computer at a data center will reduce electricity usage by 30%. Many other savings were identified during the energy-planning process described in Chapter 4, "Energy Planning."

A Gantt chart is commonly used for small projects to show who is scheduled to do what and when. An example of a Gantt chart for conducting an audit is shown in Table 5.5. Any major project should have a similar chart prepared. The chart should also include who is going to do the task.

For more complex projects, a CPM chart may be instrumental. It helps by showing exactly what activities have to be worked on every day to achieve the minimal schedule and what activities have slack so they can wait if needed.

QVS Corporation Example of Implementation

The ECMs are listed in a computer IT system and prioritized by payback period. An example of an ECM used by QVS Corporation is shown in Figure 5.5.

QVS Corporation Uses LEED

Developed by the US Green Building Council, LEED is driving green building construction today. QVS Corporation's Strategic Council, on the recommendation of the energy champion, is committed to using LEED principles and best-in-class design for any new facilities or major upgrades of existing facilities. The Strategic Council expects the following:

- Lower operating costs and increased asset value
- Conservation of energy, water, and other resources
- Healthier and safer facilities for occupants

QVS Corporation's energy champion and the energy team developed a conceptual representation of energy performance to use in their awareness training and updates to the Strategic Council. QVS Corporation's facilities have a demand for electricity and natural gas. As they enter the facility they are charged to the company by the meter, kWh consumed. SEUs use the majority of electricity, minimized where possible by good operational controls. The energy efficiencies enable O&Ts and energy action plans and projects to be developed and implemented in order to reduce electricity usage. Also, the conservation efforts of all personnel help reduce the consumption. These actions listed help determine QVS Corporation's energy performance, which is measured by the EnPIs, especially the electricity and natural gas intensity measures (see Figure 5.6).

Table 5.5 QVS Corporation's electricity audit (five facilities).

ID	Task	Start	Finish	Duration	12/9	12/16	12/23	12/30	1/6	1/13	1/20	1/27	2/3	2/10
					December 2012				January 2013				February 2013	
1	Task 1: Write statement of work (SOW)	12/7/2012	12/11/2012	3 days	▮									
2	Task 2: Obtain funds	12/7/2012	12/13/2012	1 week	▮									
3	Task 3: Negotiate with vendors	12/13/2012	12/21/2012	1 week, 2 days		▮								
4	Task 4: Select an energy auditor	12/21/2012	12/27/2012	1 week			▮							
5	Task 5: Perform audit of HQ building	1/7/2013	1/10/2013	4 days					▮					
6	Task 6: Perform audit of Plant A	1/11/2013	1/18/2013	1 week, 1 day						▮				
7	Task 7: Perform audit of Plant B	1/22/2013	1/28/2013	1 week							▮			
8	Task 8: Perform audit of Plant C	1/29/2013	2/5/2013	1 week, 2 days								▮		
9	Task 9: Perform audit of Plant D	2/7/2013	2/13/2013	1 week									▮	
10	Task 10: Write reports and discuss with energy team	2/14/2013	2/27/2013	2 weeks										▮

Energy Conservation Measure		
ECM number: Pro-11-01		
Facility and location: QVS HQ facility	Project title: Install Occupancy Controls (OC)	Date: October 2011
Project originator (name and telephone): John Steele (940) 867-3344	Estimated savings: 0.5% of total facility kWh usage	Payback period: 3.2 years
Project description/cost: Install OCs in break rooms (2), copier areas (5), mechanical rooms (4), restrooms (8), and storage rooms (12)		
Estimated savings calculations: Cost is 31 × $220 each = $6,810 with last one costing only $210. The estimated saving is 0.5% of the total electricity usage for 31 OCs.		
Payback period calculations: Cost / Savings = Years to payback = 3.2 years		

Figure 5.5 QVS Corporation ECM.

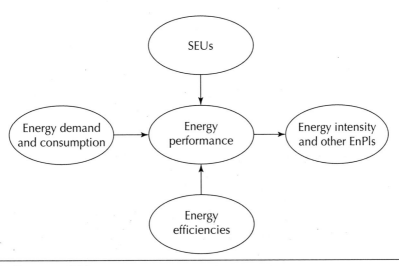

Figure 5.6 Process variables.

PROCUREMENT OF ENERGY SERVICES, PRODUCTS, AND EQUIPMENT

ISO 50001 EnMS Standard

Purpose: The organization shall establish a policy to consider energy savings and performance in all its procurement actions, such as purchasing electronics, equipment, systems, and so forth.

Procurement characteristics: The organization shall develop criteria and communicate to all management, employees, and contractors to purchase and use Energy Star electronics or equipment. The organization must define and document energy-saving practices and specifications and ensure they are communicated and used to reduce future energy consumption.

Operational Explanation

ISO 50001 EnMS encourages implementing strategies that reduce energy use. One of those strategies is to replace old equipment at the end of its life cycle with electronics, office equipment, and plant equipment that are Energy Star rated or energy efficient. Procurement needs to develop clauses to use in its purchase specifications and contracts to introduce the energy-saving or energy-efficiency requirements. QVS Corporation's procurement plan is shown later in Figure 5.7.

Forms, Templates, Processes, and Plans to Meet the Standard

Develop a plan that emphasizes the Energy Star and energy-saving features that equipment and electronics should contain if purchased. Make it applicable to all and ensure that procurement personnel adhere to the established policy.

QVS Corporation Example of Implementation

Figure 5.7 shows QVS Corporation's procurement plan.

QVS Corp. will purchase equipment that is Energy Star rated. When Energy Star is not available, then the most energy-efficient product should be purchased, providing the product is available in a reasonable time and is cost-effective. All equipment purchased should be environmentally friendly and have minimal impact on the environment. These requirements should be included in all specifications for purchasing equipment or electronics.

During operations, all electronics should be set to save energy through IT power management. At the end of life, electronics should be recycled through a responsible, sound electronics recycler, preferably an R2 electronics recycler.

All office paper should contain at least 30% recycled content. All purchased imaging equipment should have a duplex printing feature.

Critical equipment purchased will be connected by facilities or an approved contractor to the emergency power circuit (furnished by electric backup generators) in the event of a loss of electricity.

Procurement is responsible for reviewing all specifications, proposals, quotes, or qualifications to ensure the above policy is adhered to and practiced by all purchase card holders or purchasing agents.

Figure 5.7 QVS Corporation's procurement plan.

Chapter 6
Checking

The "check" phase is important to ensure that the plan or O&Ts are on track and that barriers are identified early and dealt with in order to minimize or eliminate them. For each O&T, a responsible person is assigned who will ensure that the action plans are implemented on time. At each energy team meeting, the team reviews the status of the O&T, and the operations manager checks the milestones annotated on the action plans that are due each month to ensure they are reached as planned.

The energy champion conducts a quarterly review to ensure that there are no barriers to the O&Ts and that the proper support has been provided. He or she should also brief top management at least quarterly as to the progress of and the results shown on the EnPIs. A management review should be held at least once a year that includes designated inputs and produces designated outputs that include recommended corrective and improvement actions.

MONITORING, MEASUREMENT, AND ANALYSIS
ISO 50001 EnMS Standard

Purpose: The organization must determine what should be measured and what should be calibrated, and list the frequency, method, and where documents are kept for each measured item.

Monitoring, measurement, and analysis characteristics: Measurements should include any completed O&T items that need to be measured, such as power factor readings, the percentage and number of computers and monitors in the organization's IT power management program, the percentage and number of procurement plans that include energy efficiencies in the specifications, the organization's EnPIs, evaluation of energy goals and the actual energy performance and analysis of what happened, records of equipment calibration, and the accuracy of calibrated equipment such as meters and scales. An energy plan and review should be monitored, analyzed, and updated as needed. Any deviations from the actual plan must be analyzed, corrected if possible, and documented.

Operational Explanation

As energy O&Ts or energy projects are completed, the energy team needs to ask, "Does this item need to be continuously monitored?" If so, it should be added to the monitoring and measuring plan. The organization shall identify and describe the measuring and monitoring requirements of its energy management action plans. The standard says that the organization shall ensure that the monitoring and measurement equipment related to the EnPIs provides data that are accurate and repeatable, and maintained and monitored.

Forms, Templates, Processes, CSFs, and Plans to Meet the Standard

A Microsoft Excel spreadsheet is useful in listing all the items that are measured and monitored, along with their frequency of monitoring. An example of a monitoring and measuring guide for QVS Corporation is shown later in Figure 6.3.

QVS Corporation Example of Implementation

One of the company's facilities discovered it was being charged $600 a month for a power factor adjustment fee. A check of the facility's electric bill showed its power factor ranges from 0.76 to 0.82. Texas requires a facility's power factor to be 0.95 or higher to prevent such charges (1.0 is perfect). The reason for the charge is the utility provider has to make available additional electricity that is not used because of the poor power factor from the facility. (A poor power factor may be due to the harmonics caused by all of the facility's equipment or lights being turned on at the same time rather than staggering them. The only way to increase the power factor is to add capacitors.) The facility manager brought this to the attention of the energy team. The energy team then presented this to the energy champion, who convinced the Strategic Council to fund a $16,000 project to add capacitors at the facility that would improve the power factor readings. By adding the capacitors, the company saved $600 a month, or $7,200 a year. The payback period was the cost / savings per year, or $16,000 / $7,200 = 2.22 years. In other words, in 2.22 years, the project cost will be paid for and every year thereafter will be a savings to the company. (Most company management will support a payback period of three years or less.) The energy champion asked the team to verify that an actual savings was realized. The team asked the facility manager to monitor this and to report the status to the energy team twice a year, after the capacitors were installed. The first report looked like Figures 6.1 and 6.2.

In March 2012, the facility manager presented Figures 6.1 and 6.2 to the energy team. She was thrilled that the power factor readings had all been above 0.95 and the facility was not charged a power factor adjustment fee. In fact, the savings have already paid for 22.5% of the project cost. The energy team agreed this was an excellent project. The facility manager said she plans to monitor it for three years or until total project cost payback has occurred. She looked forward to reporting back to the energy team in September 2012.

ISO 50001 EnMS states the organization shall plan and implement the monitoring, measurement, analysis, and improvement processes needed to:

- Demonstrate the performance of the EnMS

- Demonstrate the energy-efficiency performance of the organization; at planned intervals, the organization shall measure, monitor, and record significant uses that affect the EnPIs and evaluate the results

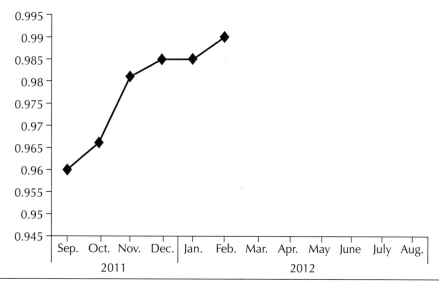

Figure 6.1 Power factor reading at QVS Corporation Plant B.

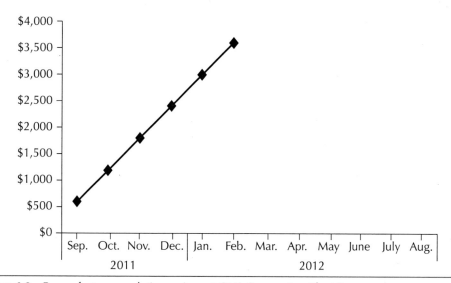

Figure 6.2 Power factor cumulative savings at QVS Corporation Plant B.

EnPIs shall be regularly monitored to measure the effectiveness of the EnMS. The indicator, what it is measuring (formula), the type of graph, who is responsible for the data collection, and the frequency of collection and graphing are listed for each EnPI in Table 6.1.

The monitoring and analysis plan is shown in Figure 6.3. Most of the EnPIs were calculated except power utilization effectiveness (PUE) for data centers. Because data centers are energy hogs, the key results performance indicators, such as server utilization, server efficiency, and PUE, should be monitored. PUE is especially important if the data center takes up most of the facility in square feet utilized. But even if it does not, it is still useful to measure any data center power

Table 6.1 EnPIs.

Indicator	Formula	Type of graph	Person responsible for:		Frequency	Reviewed by
			Data collection	Graph		
Energy intensity	kBtu/sq. ft.	Line	Energy team leader	Energy team leader	Monthly	Energy team, energy champion, Strategic Council, all company personnel
Electricity intensity, Plant A	kBtu/sq. ft.	Line	Energy team leader	Energy team leader	Monthly	Energy team, energy champion, Strategic Council, all company personnel
Natural gas intensity, Plant A	kBtu/sq. ft.	Line	Energy team leader	Energy team leader	Monthly	Energy team, energy champion, Strategic Council, all company personnel
Power factor, Plant B	Power factor reading, 0–1	Line	Facility manager, Plant B	Facility manager, Plant B	Monthly, but presented every six months	Energy team, energy champion, occasionally Strategic Council
PUE, HQ data center	Total facility load / total IT load	Column	Data center manager	Data center manager	Annually	Data center manager, HQ IT director, energy team
Greenhouse gas	Pounds of CO_2	Line	Energy team leader	Energy team leader	Annually	Energy team, energy champion, Strategic Council, all company personnel

Energy intensity, HQ and Plants A–D	kBtu/sq. ft.	Line	Facility managers	Facility managers	Monthly	Energy team, facilities staff
Electricity intensity, HQ and Plants A–D	kBtu/sq. ft.	Line	Facility managers	Facility managers	Monthly	Energy team, facilities staff
Natural gas intensity, HQ and Plants A–D	cu. ft./sq. ft.	Line	Facility managers	Facility managers	Monthly	Energy team, facilities staff
HVAC kW usage	kW and percentage of total kW used	Pie chart	Facility managers	Facility managers	Biannually	Energy team, facilities staff, energy champion
Lighting kW usage	kW and percentage of total kW used	Pie chart	Facility managers	Facility managers	Biannually	Energy team, facilities staff, energy champion
Office machine kW usage	kW and percentage of total kW used	Pie chart	Facility managers	Facility managers	Biannually	Energy team, facilities staff, energy champion
Data center/IT kW usage	kW and percentage of total kW used	Pie chart	Facility manager, HQ building	Facility manager, HQ building	Biannually	Energy team, facilities staff, energy champion, data center manager, staff IT manager

> **QVS Corporation Measuring Plan**
>
> QVS Corp.'s EnPIs (energy performance indicators) are kWh usage by month and the electricity intensity, which is kWh/total square feet. The energy team is responsible for obtaining the electric bills, recording the data monthly, and developing a bar chart with the data underneath it in a table. A corporate EnPI for the kWh usage and electricity intensity will be maintained in addition to one for each of the five facilities.
>
> **QVS Corporation Monitoring Plan**
>
> The energy team will review the EnPIs at each team meeting and furnish the objective champion with an updated copy monthly. The objective champion will review the trends, progress, and results monthly and present to the Strategic Council for their review each quarter at the Strategic Council meeting.
>
> **QVS Corporation Evaluation, Analysis, and Improvement Process**
>
> During the reviews, the energy team will determine whether the results are on track to meet the goal or target. If the results are not on track, the team will identify the gap and determine the root cause(s) for the gap. The corporate EnPIs will show the gap between projected results and actual status. The analysis will start with peeling back to the facilities to see which ones are causing the gap. Then the root causes can be identified. The team will determine what countermeasures are needed to get the EnPIs back on track. The countermeasures will include a description, the cost, the payback period, and the estimated contribution to be achieved. If countermeasures are identified with no cost, the energy team can authorize their implementation. If funds are needed, the countermeasures need to be sent to the objective champion. The objective champion will go to the Strategic Council for funds that exceed his or her approval authority.

Figure 6.3 QVS Corporation's monitoring and measurement guide.

reductions. The formula is Total Facility Load / Total IT Load. The units are in power, which is the highest kW reading for each. The normal reading is around 2.0 to 2.5; below 2.0 is desired.

Figure 6.3 shows an example of a monitoring and measuring guide for QVS Corporation.

CSFs

CSFs are those important areas, items, or factors that must happen or be achieved to make the EnMS and energy performance successful. Once the CSFs are identified, measurement of them should be developed. The CSFs are as follows:

1. Top management commitment
2. Communications
3. Sufficient contribution identified to meet targets
4. Employee involvement
5. Action plans and O&Ts
6. Resources/budget
7. Project management
8. Reviews conducted

The CSF assessment and measurement instrument is shown in Figure 6.4. Once the cumulative score of the CSFs reaches a certain level (approximately 88% of total possible score), the EnMS goes from the planning/developing and implementation stages to the maintaining and sustaining stages. At this time, an internal or external audit of the EnMS should show the organization to be in conformance with ISO 50001 EnMS-2011.

At the fourth meeting, and until QVS Corporation's energy team puts into place the objectives and projects to achieve the goal and actually achieves the goal, the CSF assessment and measurement tool (see Figure 6.4) is administered to the team members (quarterly), scored, discussed (with corrective actions taken when needed), and documented in the team meeting minutes. After the assessment and measuring tool was scored at the January 2011 meeting, the results were discussed. The 10 team members scored each of the eight CSFs. Each CSF had a maximum score of five, so the highest score possible was 40 (8×5). The team members' scores were added for each CSF and then divided by 10 to get the average score. Then, the average scores of each CSF were added to get a cumulative score for all CSFs. A graph for each CSF is recommended but is not mandatory. Once the scores are averaged, the team should review each score and ask what can be done to improve it. If it is something that can be done in a short time, the team decides whether it will accomplish this prior to the next team meeting. If so, a person(s) is assigned this responsibility.

On the first scoring of the CSF assessment and scoring tool, the scores were as shown in Table 6.2.

The team felt it was 23.3% on the way to successfully implementing an electricity reduction program and that, for now, it should continue its planned actions and develop and send to all employees a communication on the goals, champion, and energy team and ask for their ideas and support.

Every quarter the team scores the CSF assessment. Table 6.3 and Figures 6.5 and 6.6 show the 2011 and 2012 results and their trend graphs.

An increasing score is what you want. If the energy team is engaged, then you will probably get a higher score. However, if a project or objective fails or a key person is no longer available, it is possible for the overall CSF score to decrease. The last total CSF score was 29.9 out of 40 possible, or 74.8%. It has been the author's experience that once you achieve 88%–93%, the implementation stage becomes the maintenance stage. How long should you keep administering the assessment tool to the energy team? You should continue as long as it is helpful. Let's look at December 2012 in Table 6.3 and see whether there is any useful info to help us improve. First, top management has a score of 4.2, which is good. Its continued involvement should be encouraged by the objective champion. Communications is at 3.5. The communications plan was recently completed. As it is fully implemented over time and several communications media are employed, this CSF will increase. Sufficient contribution at 3.7 and resources at 3.5 should be evaluated by the energy team prior to the January 25, 2013, management review so that any additional resources needed for projects can be requested and discussed. Employee involvement at 4.1 shows the employees support the energy conservation program. The objectives/action plans and the project plans at 3.7 and 3.5, respectively, should increase after the resources-needed review is made. If few resources are needed, then both of these are okay. If more resources are needed, then both of these will need work. The reviews at 2.2 are the lowest, but

Review each of the eight CSFs and circle the most current and appropriate answer for each. The facilitator will then collect your score for each CSF and your total score. The minimum possible score is 8 and the maximum is 40 (8 CSFs × 5).

1. **Top Management Commitment (1–5)**
   1. No commitment shown.
   2. Formed a headquarters energy team and wants a QVS Corporation energy plan developed.
   3. Desires to meet all ISO 50001 EnMS requirements and willing to provide resources to accomplish. Has an energy policy and has appointed an energy management representative.
   4. Shows funding support on key energy reduction initiatives and projects.
   5. "Talks the talk and walks the walk" all the way to meeting established energy goals. Conducts a management review annually to ensure suitability and effectiveness of the EnMS. Inputs and outputs to the management reviews were as required.

2. **Communications (1–5)**
   1. No specific communications on reducing energy to QVS employees/contractors.
   2. Some communication of energy goals and QVS's intent to achieve them.
   3. Communications on specific items or activities that every employee and contractor can do to help reduce energy use have been written and communicated.
   4. All projects, conservation measures, and other energy reduction countermeasures undertaken by the facility and progress toward meeting goals are communicated on a periodic basis.
   5. All employees and management know the QVS energy plan, are making progress toward achieving EnMS goals, and are actively participating in reducing energy use.

3. **Sufficient Contribution Identified to Meet Target (1–5)**
   1. No energy reduction projects have been identified, and no significant energy conservation program has been implemented.
   2. Projects have been identified, but no estimate of how many kWh of electricity they will reduce has been accomplished.
   3. Energy audits have been conducted and energy projects have been identified with a rate of return, but none or only a small number have been funded.
   4. Adequate bundling of projects enabling identification of sufficient reductions to meet project-related targets and pay for project costs through savings realized or QVS HQ has funded a significant number of projects that, when completed, will achieve established targets.
   5. Project-related targets are met and adequate savings are available to pay for all utility energy services contracts/energy savings performance contracts (UESC/ESPC) projects, if applicable. An energy conservation program is in effect to help maintain the reductions and savings in electricity costs.

4. **Employee Involvement (1–5)**
   1. Most not aware of the EnMS, corporate energy policy, and operational controls that may apply to them.
   2. Only those with energy-related jobs are involved. Corporation has not communicated operational controls or operational energy criteria where needed.
   3. A large number of employees are involved in energy conservation efforts, especially electricity conservation. Operational controls have been identified and communicated, and training is provided if needed.
   4. Employees are aware of and supportive of electricity reduction targets and are fully engaged to help wherever possible. Operational controls are in place and employees are aware of and are following them.

Figure 6.4 CSF measurement and assessment tool.

5. Almost everyone is involved in some electricity conservation and/or use-reduction efforts. Electricity reductions are occurring and continual improvements have been achieved. Everyone follows guidelines, and processes are performed safely and with excellent results. Electricity reduction is everyone's business. World-class performance is happening.

5. **Action Planning: Objectives and Targets (1–5)**
   1. None developed.
   2. Accomplished energy profile, identified legal requirements, and currently developing some objectives.
   3. High-energy sources have been identified and objectives are being developed. Action plans are being developed for each objective. The objectives, and specifically the targets, are SMART (specific, measurable, actionable, relevant, and time framed).
   4. Action plans have been developed, responsibilities have been outlined, resources have been approved, and implementation of improvement actions has begun. A management review has been conducted and objectives have been approved. A contingency plan for achieving emergency electricity has been written and approved.
   5. Some or all of the objectives and targets have been achieved, and new ones with clear and complete action plans are being developed.

6. **Resources/Budget (1–5)**
   1. No resources provided except appointing the energy team and providing a meeting place and time.
   2. Resources have been requested to achieve objectives and targets. Analysis has been performed to identify and measure high-energy users. A measurement system is in place, tying funds provided to results achieved.
   3. Some funds/resources have been provided. Approved first by the objective champion and Strategic Council.
   4. Sufficient resources are available for all action plans and projects.
   5. Separate budget with ample funds is in place.

7. **Project Management (1–5)**
   1. No project management evident even with objectives deployed.
   2. Projects have been reviewed, statements of work (SOWs) have been developed, and contract advertisement and award is in process. Improving energy efficiency in design is evident.
   3. A project management system is now in place, and projects are being reviewed by project managers for schedule, cost, and quality. Objectives implementation is being reviewed by energy team for progress and to identify and correct any barriers to progress.
   4. Projects and objectives are being reviewed and managed.
   5. Some projects and objectives have been completed with good results, and others are progressing well.

8. **Reviews (1–5)**
   1. No team reviews, project reviews, or management reviews have occurred.
   2. Projects have been developed and implemented, and both project and team reviews have been scheduled.
   3. A project review has been accomplished, and the energy team has performed a team review at least once per month.
   4. A management review has been scheduled, and numerous team reviews have been conducted.
   5. A management review has been conducted by senior management, corrective actions have been taken where needed, numerous project and team reviews have been conducted, and projects are on track.

Figure 6.4 CSF measurement and assessment tool. *(Continued)*

Table 6.2 First use of CSF assessment tool (March 2011).

Individual CSF	Team average	Actions needed	Person(s) responsible
Top management	2.2	None at this time	
Communications	1.1	Need communications including goals, energy team formed	Team leader in coordination with champion
Sufficient contribution estimated	1.0	On track	
Employee involvement	1.0	None at this time	
Resources	1.0	None at this time	
Objectives/action plans	1.0	None at this time	
Project plans	1.0	None at this time	
Reviews	1.0		
Total	9.3		

Table 6.3 CSF assessments.

Individual CSF	March 2011	June 2011	September 2011	December 2011
Top management	2.2	2.3	2.3	2.5
Communications	1.1	1.1	1.3	1.3
Sufficient contribution estimated	1.0	2.5	3.0	3.0
Employee involvement	1.0	1.0	2.4	2.7
Resources	1.0	1.0	2.0	2.5
Objectives/action plans	1.0	1.5	1.7	2.3
Project plans	1.0	1.2	1.4	1.5
Reviews	1.0	1.0	1.0	1.3
Total	9.3	11.6	15.1	17.1

Table 6.3 CSF assessments. *(Continued)*

Individual CSF	March 2012	June 2012	September 2012	December 2012
Top management	2.6	2.9	3.5	4.2
Communications	2.0	2.5	2.8	3.5
Sufficient contribution estimated	3.1	3.5	3.6	3.7
Employee involvement	3.1	3.3	3.9	4.1
Resources	2.7	3.0	3.3	3.5
Objectives/action plans	2.6	3.1	3.3	3.7
Project plans	2.6.	2.9	3.1	3.5
Reviews	1.4	1.8	2.0	2.2
Total	20.1	23.0	25.5	29.9

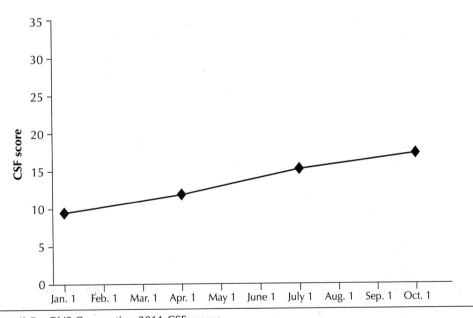

Figure 6.5 QVS Corporation 2011 CSF scores.

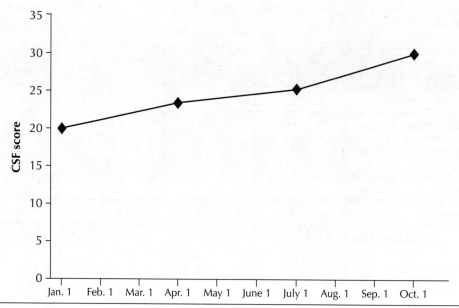

Figure 6.6 QVS Corporation 2012 CSF scores.

this should drastically increase when the January 25, 2013, management review is accomplished. It is obvious that these scores are helpful now, and the assessment should be administered again on schedule at the end of March 2013.

An interesting use of the CSF assessment tool is to have the objective champion and the Strategic Council members take the assessment themselves. It will be helpful to hear why they scored certain areas the way they did. Also, a comparison between the council and the energy team should prove beneficial to both and help them better understand the actual progress that has resulted.

The document control manager or the team leader should maintain graphs over time showing how the individual CSFs are scored. The objective champion will want to see how the top management, employee involvement, and resources CSFs are doing.

It is useful to graph an individual CSF and have the team identify what it takes to increase the CSF's score in the near future and in the longer term. That course of action increases both the CSF score and team progress on the O&Ts.

EVALUATION OF COMPLIANCE WITH LEGAL REQUIREMENTS

ISO 50001 EnMS Standard

Purpose: The energy team must periodically evaluate how well the legal requirements and other requirements are being complied with, analyze the root causes of any deviation, and take corrective action.

Legal requirements compliance characteristics: These evaluations should be provided to the management representative, presented as an input to the management reviews, and documented in the organization's document control system. The interval of evaluation is not specified, but once a year is acceptable.

Operational Explanation

Once a year, the energy team reviews the list of legal requirements. One by one, the team asks and answers the following questions:

1. Do we comply with this legal requirement? If not, why?
2. Are we in complete compliance or partial compliance?
3. What do we need to do to be in compliance?

Forms, Templates, Processes, and Plans to Meet the Standard

Document the legal and other requirements annual review. Make a copy of the list and annotate whether the organization is in compliance with each requirement. Have the team leader sign and date the list. Place it in the document control system in the folder titled "Evaluation of compliance with legal and other requirements."

QVS Corporation Example of Implementation

Table 6.4 and Figure 6.7 show QVS Corporation's legal and other requirements compliance evaluations.

Table 6.4 Legal and other requirements compliance evaluation, 2012 (short form).

Requirement	In compliance? (yes/no)	Remarks
1. Purchase of electricity	Yes	Negotiated an electricity provider agreement with Green Mountain Energy Company, reducing QVS Corp.'s cents per kWh by 25% while getting 100% pollution-free electricity from the wind
2. Design and operations: ASHRAE and IEEE standards	Yes	The engineers' designs adhered to ASHRAE and IEEE standards
3. Electricity at Work requirements	Yes	Inspection showed all electrical items in great shape
4. EPA ventilation requirements	Yes	All ventilation requirements were properly followed
5. County and community building codes	Yes	All county codes for building installation to include electricity and natural gas were properly followed
6. State requirements for power factor	Yes	One plant was having power factor problems. The energy team had capacitors put into place that negated the problem.
7. National Environmental Policy Act and other federal legislation	Yes	In compliance with all pertinent legislation

> The purchase of electricity is determined by utility rates and considerations such as cents per kWh and other fees and adjustments such as power factor adjustment. This will be spelled out in the agreement with the utility provider and is normally included in the electric bill. It may also be impacted by Texas Senate Bill 7, passed January 1, 2002, regarding electricity deregulation in Texas.
>
> *Evaluation comments:* Negotiated an electricity provider agreement with Green Mountain Energy Company, reducing QVS Corp.'s cents per kWh by 25% while getting 100% pollution-free electricity from the wind.
>
> Design and operations are determined by consideration of the following standards and legislation:
>
> - ASHRAE (American Society of Heating, Refrigerating, and Air-Conditioning Engineers) standards for heating, air-conditioning, ventilation, and data centers (3, 15, 62, and 90.1), consisting of three types of standards:
>   1. Methods of testing
>   2. Standard design
>   3. Standard practices
> - ASME (American Society of Mechanical Engineers) and IEEE (Institute of Electrical and Electronics Engineers) standards on insulation, high-voltage testing, and hooking items to electronic equipment
>
>   *Evaluation comments:* The engineers' designs adhered to ASHRAE and IEEE standards.
> - Federal legislation including the National Environmental Policy Act (NEPA), the Natural Gas Policy Act (NGPA), and the Public Utility Regulatory Policies Act (PURPA)
>
>   *Evaluation comments:* All pertinent legislation is followed where applicable.
> - Electricity at Work regulations that require companies to maintain all electrical items and systems in a safe and workable condition
>
>   *Evaluation comments:* Inspection showed all electrical items in great shape.
> - Environmental Protection Agency (EPA) regulations for meeting ventilation requirements in offices
>
>   *Evaluation comments:* All ventilation requirements were properly followed.
> - County and community statutes and codes for building installation to include electricity and natural gas requirements
>
>   *Evaluation comments:* All county codes for building installation to include electricity and natural gas were properly followed.
> - State requirements for power factor and other issues
>
>   *Evaluation comments:* One plant was having power factor problems. The energy team had capacitors put into place that negated the problem.
>
> **Compliance Evaluation**
> Compliance evaluation of the legal and other requirements will be added to the QVS Corp.'s internal audit or self-inspection checklist and evaluated as part of the annual self-inspection. Documentation of the results, including any corrective action efforts, will be maintained.
>
> 1. Has each facility been evaluated for electricity legal requirements and are they documented? ☒ Yes ☐ No
> 2. Have specific energy laws been identified and documented? ☒ Yes ☐ No

Figure 6.7 Legal and other requirements compliance evaluation, 2012 (long form).

3. Have the legal requirements been reviewed within the past year? ☒ Yes ☐ No

   (Date of last review: _____)

4. Is there evidence that the facility has evaluated its compliance with the electricity and natural gas legal requirements? ☒ Yes ☐ No

5. If the facility has aspects that are not in compliance with applicable legal requirements, have corrective actions been taken? ☒ Yes ☐ No

   Comments: _QVS Corp. is in compliance with its legal and other requirements._

Figure 6.7 Legal and other requirements compliance evaluation, 2012 (long form). *(Continued)*

INTERNAL AUDIT OF EnMS

ISO 50001 EnMS Standard

Purpose: The organization shall perform periodically at a planned interval (every year is acceptable) an internal audit to ensure its EnMS conforms to the ISO standard requirements.

Internal audit characteristics: The internal audit needs to be conducted by a self-inspection by the energy team, the management representative, or the organization's audit team or department. It should evaluate how well the team is meeting the standard's elements, including how aware people are of the EnMS and the energy policy; how effective the energy O&Ts have been; how well the energy action plans spell out all the tasks that need to be done to achieve the targets; how effective are the meetings, management reviews, and the monitoring and measurement tracking and analysis; and how the energy performance changed and why. The energy team can do self-inspections, and other cross-functional personnel may be selected to conduct an audit. In the latter case, the management representative would train them and approve the audit plan. Records of self-inspections or internal audits must be maintained and presented at the appropriate management review.

Operational Explanation

An internal audit or self-inspection should be conducted at least every 12 months to identify any nonconformances or deficiencies. A self-inspection checklist should be developed and filled out through research or other team actions and discussion annually. Any major deficiencies noted should require a CAR that identifies the problem, finds the root cause, and solves the problem in a timely manner. It also provides the correct information for documenting the problem and its resolution.

The self-inspection checklist enables an internal audit of actual achievements versus standards requirements in the following 12 major areas:

1. Roles and responsibilities
2. Energy policy
3. SEUs
4. Legal and other requirements
5. O&Ts and their action plans
6. Training
7. Communications
8. Documentation and document control
9. Operational controls
10. Monitoring and measurement
11. Facility auditing and corrective actions
12. Management reviews

This same checklist can be used by an organization's headquarters audit team in performing an audit of a facility or an overall organizational audit. The internal audit can be done by the headquarters internal audit department provided it develops an audit plan and assigns auditors who have had some energy management training. This second-party audit needs to be done every three years. The audit will consist of using the self-inspection checklist and interviewing personnel from top management, employees to determine their awareness, the energy champion, the energy team members, the facility managers, and whomever else the audit team wants to interview. They will confirm that what the EnMS implementation plan said it was going to do has actually been done. A check on the EnPIs will show the progress and results that have been achieved.

Forms, Templates, Processes, Checklists, and Plans to Meet the Standard

Figure 6.8 shows a self-inspection checklist.

QVS Corporation Example of Implementation

The energy team completed its first self-inspection on September 29, 2011 (Figure 6.9).

Element 1: Roles and responsibilities

1. Have energy team member roles been identified and documented in team meeting minutes and reported to the objective champion?   ☐ Yes   ☐ No

2. Is the team worksheet current?   ☐ Yes   ☐ No
   (Date of last review: _____)

3. Do team members with identified roles understand their responsibilities and are they being fulfilled?   ☐ Yes   ☐ No

---

Figure 6.8   Sample internal audit/self-inspection checklist.

4. Do EnMS team members participate in meetings? ☐ Yes ☐ No

5. Are facility staff and management aware of their responsibilities with regard to the EnMS? ☐ Yes ☐ No

Comments: _____

_____

_____

Element 2: Energy policy

6. Does the organization have an EnMS policy? ☐ Yes ☐ No

7. Has the EnMS policy been approved by the senior leadership? ☐ Yes ☐ No

8. Is the policy current and reviewed annually? ☐ Yes ☐ No
   (Date of last review: _____)

9. Has the policy been provided and communicated to employees and contractors? ☐ Yes ☐ No

Comments: _____

_____

_____

Element 3: SEUs

10. Have the SEUs for each facility been identified? ☐ Yes ☐ No

11. Has the percentage of each SEU been estimated? ☐ Yes ☐ No

12. Have the SEUs been evaluated as to reduction possibilities? ☐ Yes ☐ No

13. Have the SEUs been prioritized? ☐ Yes ☐ No

14. Have relevant energy variables been identified and energy efficiencies identified and documented? ☐ Yes ☐ No

15. Have the SEUs been reviewed within the past year? ☐ Yes ☐ No
    (Date of last review: _____)

Comments: _____

_____

_____

Element 4: Legal and other requirements

16. Has each facility been evaluated for electricity and natural gas legal requirements and are they documented? ☐ Yes ☐ No

17. Have specific energy laws been identified and documented? ☐ Yes ☐ No

18. Have the legal requirements been reviewed within the past year? ☐ Yes ☐ No
    (Date of last review: _____)

19. Is there evidence that the facility has evaluated its compliance with the electricity and natural gas legal requirements? ☐ Yes ☐ No

Figure 6.8 Sample internal audit/self-inspection checklist. *(Continued)*

(continued)

20. If the facility has aspects that are not in compliance with applicable legal requirements, have corrective actions been taken? ☐ Yes ☐ No

Comments: _____

_____

_____

Element 5: Objectives, targets, and environmental action plans (EAPs)

21. Has the energy team worked on or initiated at least one objective and target this year? ☐ Yes ☐ No

22. Have objectives, targets, and EAPs been documented? ☐ Yes ☐ No

23. Do objectives and targets address the respective facility SEUs and energy efficiencies? ☐ Yes ☐ No

24. Has a documented action plan been put in place for each objective designating responsibility, detailing steps to completion, and describing how the team will successfully meet the target? ☐ Yes ☐ No

25. Are objectives being reviewed for progress at the team meetings? ☐ Yes ☐ No

26. Is the status for each action plan action reflected using a stoplight symbol at each meeting? ☐ Yes ☐ No

(Date of last review: _____)

Comments: _____

_____

_____

Element 6: Training

27. Has energy awareness training been developed and presented to employees and contractors? ☐ Yes ☐ No

28. Have the training needs been reviewed within the past year? ☐ Yes ☐ No

(Date of last review: _____)

29. Have employees who require training been trained and is the training current? ☐ Yes ☐ No

30. Did the facility and competency training meet the ISO 50001 EnMS requirements? ☐ Yes ☐ No

Comments: _____

_____

_____

Element 7: Communication

31. Do the objective champion and energy team communicate relevant information to the facility employees and contractors? ☐ Yes ☐ No

32. Do employees know to whom to communicate energy concerns? ☐ Yes ☐ No

Figure 6.8 Sample internal audit/self-inspection checklist. *(Continued)*

33. Are external communications adequately addressed? ☐ Yes ☐ No
Comments: _____
_____
_____

Element 8: Operational controls

34. Are SEUs with legal requirements addressed with an operational control, an objective and target, or a project? ☐ Yes ☐ No
35. Have operational controls been identified and briefed to those impacted? ☐ Yes ☐ No
36. Do all SEUs have at least one operational control? ☐ Yes ☐ No
37. Is there evidence that operational controls are functional and that they adequately address the SEUs? ☐ Yes ☐ No
38. Have the operational controls been reviewed within the past year? ☐ Yes ☐ No
    (Date of last review: _____)
Comments: _____
_____
_____

Element 9: Documentation and control of documents

39. Is all required documentation filed on the facility's EnMS document control site? ☐ Yes ☐ No
40. Is all documentation filed in the correct folders according to the energy documentation guide? ☐ Yes ☐ No
41. Is all documentation on the document control site current? ☐ Yes ☐ No
42. Are the documents used by the energy team the current QVS Corp. EnMS program documents? ☐ Yes ☐ No
43. Does the EnMS documentation provide direction and instructions to other related documents, records, reports, schedules, and registers? ☐ Yes ☐ No
Comments: _____
_____
_____

Element 10: Monitoring and measurement

44. Have SEUs that require monitoring or measurement been identified on the monitoring and measurement plan? ☐ Yes ☐ No
45. Has the monitoring and measurement plan been reviewed within the past year? ☐ Yes ☐ No
    (Date of last review: _____)
46. Is the required monitoring and measurement being conducted for the energy performance indicators? ☐ Yes ☐ No

Figure 6.8 Sample internal audit/self-inspection checklist. *(Continued)*   *(continued)*

47. Is the monitoring and measurement relevant to the SEUs?	☐ Yes	☐ No
48. Is any necessary calibration of equipment included in the monitoring plan?	☐ Yes	☐ No

Comments: _____
_____
_____

Element 11: Facility auditing and corrective action

49. Have self-inspections been completed annually?	☐ Yes	☐ No
50. Were all nonconformances identified and was corrective action taken to address each nonconformance? (Date of last review: _____)	☐ Yes	☐ No
51. Were all CARs from the previous second-party audits resolved appropriately?	☐ Yes	☐ No

Comments: _____
_____
_____

Element 12: Management review

52. Has the management review been conducted at least annually? (Describe the management review process and date(s) of last review: _____)	☐ Yes	☐ No
53. Is the head of the facility, or his/her designee, aware of the status of the EnMS program?	☐ Yes	☐ No
54. Does the head of the facility, or his/her designee, provide guidance and direction for the EnMS?	☐ Yes	☐ No
55. Does the management review cover the required elements including both inputs and outputs?	☐ Yes	☐ No

Comments: _____
_____
_____

Self-inspection and nonconformances

Document all nonconformances identified in the self-inspection and the corrective action taken.

_____
_____
_____
_____
_____
_____

Figure 6.8 Sample internal audit/self-inspection checklist. *(Continued)*

Element 1: Roles and responsibilities

1. Have energy team member roles been identified and documented in team meeting minutes and reported to the objective champion? ☒ Yes ☐ No
2. Is the team worksheet current? ☐ Yes ☒ No
   (Date of last review: _May 2011_)
3. Do team members with identified roles understand their responsibilities and are they being fulfilled? ☒ Yes ☐ No
4. Do EnMS team members participate in meetings? ☒ Yes ☐ No
5. Are facility staff and management aware of their responsibilities with regard to the EnMS? ☒ Yes ☐ No

Comments: _____
1, 3, 4, & 5 were checked yes. No team worksheet had been developed so 2 was marked no.

Element 2: Energy policy

6. Does the organization have an EnMS policy? ☒ Yes ☐ No
7. Has the EnMS policy been approved by the senior leadership? ☒ Yes ☐ No
8. Is the policy current and reviewed annually? ☒ Yes ☐ No
   (Date of last review: _May 2011_)
9. Has the policy been provided and communicated to employees and contractors? ☒ Yes ☐ No

Comments: _____
            6–9 were checked yes. No nonconformity.

Element 3: SEUs

10. Have the SEUs for each facility been identified? ☒ Yes ☐ No
11. Has the percentage of each SEU been estimated? ☒ Yes ☐ No
12. Have the SEUs been evaluated as to reduction possibilities? ☒ Yes ☐ No
13. Have the SEUs been prioritized? ☒ Yes ☐ No
14. Have relevant energy variables been identified and energy efficiencies identified and documented? ☒ Yes ☐ No
15. Have the SEUs been reviewed within the past year? ☒ Yes ☐ No
    (Date of last review: _May 2011_)

Comments: _____
            10–15 were checked yes. No discrepancies.

Figure 6.9 QVS Corporation EnMS internal audit/self-inspection checklist.   (*continued*)

Element 4: Legal and other requirements

16. Has each facility been evaluated for electricity and natural gas legal requirements and are they documented? ☒ Yes ☐ No
17. Have specific energy laws been identified and documented? ☒ Yes ☐ No
18. Have the legal requirements been reviewed within the past year? ☒ Yes ☐ No
    (Date of last review: _May 2011_)
19. Is there evidence that the facility has evaluated its compliance with the electricity and natural gas legal requirements? ☒ Yes ☐ No
20. If the facility has aspects that are not in compliance with applicable legal requirements, have corrective actions been taken? ☒ Yes ☐ No

Comments: _____
16–20 were checked yes. No discrepancies.
_____

Element 5: Objectives, targets, and environmental action plans (EAPs)

21. Has the energy team worked on or initiated at least one objective and target this year? ☒ Yes ☐ No
22. Have objectives, targets, and EAPs been documented? ☒ Yes ☐ No
23. Do objectives and targets address the respective facility SEUs and energy efficiencies? ☒ Yes ☐ No
24. Has a documented action plan been put in place for each objective designating responsibility, detailing steps to completion, and describing how the team will successfully meet the target? ☒ Yes ☐ No
25. Are objectives being reviewed for progress at the team meetings? ☒ Yes ☐ No
26. Is the status for each action plan action reflected using a stoplight symbol at each meeting? ☒ Yes ☐ No
    (Date of last review: _May 2011_)

Comments: _____
21–26 were checked yes. No deficiencies.
_____

Element 6: Training

27. Has energy awareness training been developed and presented to employees and contractors? ☒ Yes ☐ No
28. Have the training needs been reviewed within the past year? ☒ Yes ☐ No
    (Date of last review: _May 2011_)
29. Have employees who require training been trained and is the training current? ☒ Yes ☐ No
30. Did the facility and competency training meet the ISO 50001 EnMS requirements? ☒ Yes ☐ No

Figure 6.9 QVS Corporation EnMS internal audit/self-inspection checklist. *(Continued)*

Comments: _____
                27–30 were checked yes. No nonconformities.
_____

Element 7: Communication

31. Do the objective champion and energy team communicate relevant information to the facility employees and contractors?   ☒ Yes   ☐ No
32. Do employees know to whom to communicate energy concerns?   ☒ Yes   ☐ No
33. Are external communications adequately addressed?   ☒ Yes   ☐ No

Comments: _____
                31–33 were checked yes. No nonconformities.
_____

Element 8: Operational controls

34. Are SEUs with legal requirements addressed with an operational control, an objective and target, or a project?   ☒ Yes   ☐ No
35. Have operational controls been identified and briefed to those impacted?   ☒ Yes   ☐ No
36. Do all SEUs have at least one operational control?   ☒ Yes   ☐ No
37. Is there evidence that operational controls are functional and that they adequately address the SEUs?   ☒ Yes   ☐ No
38. Have the operational controls been reviewed within the past year?   ☒ Yes   ☐ No
    (Date of last review: _May 2011_)

Comments: _____
                34–38 were checked yes. No nonconformities.
_____

Element 9: Documentation and control of documents

39. Is all required documentation filed on the facility's EnMS document control site?   ☒ Yes   ☐ No
40. Is all documentation filed in the correct folders according to the energy documentation guide?   ☒ Yes   ☐ No
41. Is all documentation on the document control site current?   ☐ Yes   ☒ No
42. Are the documents used by the energy team the current QVS Corp. EnMS program documents?   ☒ Yes   ☐ No
43. Does the EnMS documentation provide direction and instructions to other related documents, records, reports, schedules, and registers?   ☒ Yes   ☐ No

Comments: _____
        39, 40, 42, and 43 were checked yes. 41c was checked no, a deficiency.
_____

Figure 6.9   QVS Corporation EnMS internal audit/self-inspection checklist.   *(continued)*
(*Continued*)

Element 10: Monitoring and measurement

44. Have SEUs that require monitoring or measurement been identified on the monitoring and measurement plan? ☒ Yes ☐ No

45. Has the monitoring and measurement plan been reviewed within the past year? ☒ Yes ☐ No
    (Date of last review: _May 2011_)

46. Is the required monitoring and measurement being conducted for the energy performance indicators? ☒ Yes ☐ No

47. Is the monitoring and measurement relevant to the SEUs? ☒ Yes ☐ No

48. Is any necessary calibration of equipment included in the monitoring plan? ☒ Yes ☐ No

Comments: _____
44–48 were checked yes. No deficiencies.

Element 11: Facility auditing and corrective action

49. Have self-inspections been completed annually? ☐ Yes ☐ No

50. Were all nonconformances identified and was corrective action taken to address each nonconformance? ☒ Yes ☐ No
    (Date of last review: _May 2011_)

51. Were all CARs from the previous second-party audits resolved appropriately? ☐ Yes ☐ No

Comments: _____
49. This was the first self inspection required. 50. Yes, two deficiencies identified. 51. N/A

Element 12: Management review

52. Has the management review been conducted at least annually? ☒ Yes ☐ No
    (Describe the management review process and date(s) of last review: _May 2011_)

53. Is the head of the facility, or his/her designee, aware of the status of the EnMS program? ☒ Yes ☐ No

54. Does the head of the facility, or his/her designee, provide guidance and direction for the EnMS? ☒ Yes ☐ No

55. Does the management review cover the required elements including both inputs and outputs? ☒ Yes ☐ No

Comments: _____
52–55 were checked yes. No deficiencies.

Figure 6.9 QVS Corporation EnMS internal audit/self-inspection checklist. *(Continued)*

Self-inspection and nonconformances

Document all nonconformances identified in the self-inspection and the corrective action taken.

No team profile or worksheet developed and all documents are not included in the document control system that is required.

Figure 6.9 QVS Corporation EnMS internal audit/self-inspection checklist. *(Continued)*

NONCONFORMITIES, CORRECTIVE ACTIONS, AND PREVENTIVE ACTIONS

ISO 50001 EnMS Standard

Purpose: To correct or prevent any nonconformities or deficiencies.

Characteristics: Once the energy team identifies through either a self-inspection or a second-party audit that a nonconformity exists or will exist if preventive action is not taken, a CAR or PAR (whichever is appropriate) is developed, corrective action is taken, and correction of the nonconformity is verified.

Operational Explanation

When a discrepancy or a nonconformance to a requirement is found, action in the form of a CAR or PAR, depending on whether it is something that needs immediate correction or is something that is preventable, needs to be taken immediately. The CAR or PAR is designed to show the problem, the root causes, and whether a fix was implemented and is adequate.

Forms, Templates, Processes, and Plans to Meet the Standard

Figure 6.10 shows a typical CAR.

QVS Corporation Example of Implementation

Two deficiencies were identified during the self-inspection: (1) there was no energy team profile worksheet and (2) some documents had not been posted in the document control system. The energy team profile worksheet could be completed after the self-inspection in about 20 minutes (see Figure 6.11). Therefore, the energy team leader stated, "Let's prepare the profile sheet now; thus a CAR will not be necessary. We will need to prepare a CAR for the deficiency of the missing documents" (see Figure 6.12). Figure 6.13 shows QVS Corporation's nonconformities and corrective and preventive action process.

Corrective Action Request				
CAR # (start with En):		Location:	Policy reference:	Date issued:
Requirement:				
Nonconformance:				
Corrective action issued by:		Corrective action assigned to:		
^		Corrective action due by:		
Cause analysis EnMS team meeting date (if applicable): Root cause analysis:				
Corrective action Response: Supporting documentation:				
Acceptance Additional audits required? ☐ Yes ☐ No Corrective action accepted? ☐ Yes ☐ No Associated preventive action request (PAR) if applicable EnMS team leader _____ Date accepted _____ Office head _____ Date accepted _____				

Figure 6.10 Typical CAR.

Source: US Drug Enforcement Agency, EMS-09-014 Nonconformity, Corrective, and Preventative Action (09/22/2009).

Instructions for preparing a CAR

Block 1 of form: Description of nonconforming work or departure from policies and procedures in the Energy Management System (EnMS) or energy regulations. This section records the following information:

a. CAR #: The name of the facility followed by "EnMS CAR," the year, and a sequential numbering system (e.g., Plant A EnMS CAR 12-001).
b. Policy reference: Enter the appropriate reference to a policy, procedure, or clause addressing the nonconformance.
c. Date issued: Enter the date that the CAR was issued.
d. Requirement: Enter a brief description of the requirement cited in the policy reference.
e. Nonconformance: Enter a detailed description of the nonconformance.
f. Corrective action issued by: Enter the name and title of the person issuing the CAR.
g. Corrective action assigned to: Enter the name of the individual assigned the corrective action.
h. Corrective action due by: Enter the deadline for the corrective action.

Block 2 of form: Cause analysis is the most important and sometimes the most difficult part in the corrective action process. The procedure for corrective action shall start with an investigation to determine the root cause(s) of the nonconformance. If necessary, the EnMS team should convene, discuss the nonconformance, and determine the root cause of the nonconformance. Potential root causes could include policies, methods and procedures, staff skills and training, and equipment. This section shall record a detailed description of the root cause analysis, while including any appropriate references and attachments.

Block 3 of form: Corrective actions are the steps implemented by the respective office to correct the nonconformance and prevent its recurrence. Corrective actions shall be appropriate to the magnitude and the risk of the nonconformance. This section shall record a detailed description of the corrective action chosen by the respective office. In addition, this section shall include all supporting documentation.

Block 4 of form: All documentation will be provided to the EnMS team leader for review. The EnMS team leader will determine whether further action is required. If no further action is required, the EnMS team leader will coordinate with the objective champion for concurrence. This section will be completed as follows:

a. Additional audits required: Check either "yes" or "no" to indicate to answer whether additional audits are required. If "yes," provide reference to additional audit documentation.
b. Corrective action accepted: Check either "yes" or "no" to indicate whether the EnMS team leader and the objective champion have accepted the corrective action. If no corrective action is necessary, enter "yes" and indicate "N/A" with a brief description under the corrective action section.
c. Associated PAR: Enter the PAR number associated with the CAR (if applicable). The PAR can use this form by just writing PAR over the CAR heading.
d. Signatures and dates: Pending complete concurrence, the EnMS team leader and the objective champion will sign and date the CAR. By signing, both the EnMS team leader and the objective champion agree with all determinations (i.e., cause analysis, corrective action, etc.) and render the CAR complete.

Figure 6.10 Typical CAR. *(Continued)*

Date formed: September 25, 2010	Head of organization: Gene Smith	Team leader: Bill Johnson (817) 534-6678 bj@qvsc.com	Team facilitator: Marv Howell (817) 534-6682 mh@qvsc.com

Team members		
Name	**Telephone**	**E-mail**
Ann Docum	(817) 534-6690	ad@qvsc.com
Steve Thomas	(817) 534-6691	st@qvsc.com
John Jones	(817) 534-6692	jj@qvsc.com
Mary Ray	(817) 534-6695	mr@qvsc.com
Bill Davis	(817) 534-6700	bd@qvsc.com
Ted Pruitt	(817) 534-6707	tp@qvsc.com
John Williams	(817) 534-2012	jw@qvsc.com
Mary Walker	(817) 534-2875	mw@qvsc.com

Document control manager		
Mary Ray	(817) 534-6695	mr@qvsc.com

Operations manager		
Bill Davis	(817) 534-6700	bd@qvsc.com

Note taker		
John Jones	(817) 534-6692	jj@qvsc.com

Approved

Truman Johnson, September 2010
EnMS team leader

Figure 6.11 QVS Corporation energy team profile worksheet.

Corrective Action Request			
CAR #: CAR #1	Location: Gun Barrel City	Policy reference: N/A	Date issued: 9/25/2010
Requirement: Document Control ISO 50001 (EnMS Requirement 4.5.4)			
Nonconformance: Some documents missing from the document control system.			

Figure 6.12 QVS Corporation completed CAR.

Corrective action issued by: Truman Johnson	Corrective action assigned to: Mary Walker
	Corrective action due by: 11/30/2010

Cause analysis

EnMS team meeting date (if applicable): 10/2010

Root cause analysis:

Not checking the system files to ensure current documents have been placed.

Corrective action

Response:

The document control manager will check the files monthly to ensure all documents are current. Meeting minutes and sign-in sheets, which accounted for 70% of the missing documents, will be placed in the files no later than three days after the meeting. The energy team leader will check the files four days after the meeting to verify that the documents have been filed.

Supporting documentation:

Not needed.

Acceptance

Additional audits required? ☐ Yes ☒ No

Corrective action accepted? ☒ Yes ☐ No

Associated preventive action request (PAR) if applicable: N/A

EnMS team leader _____Truman Johnson_____ Date accepted __10/5/2010__

Office head _____Gene Smith_____ Date accepted __10/6/2010__

Figure 6.12 QVS Corporation completed CAR. *(Continued)*

During any self-inspection, any major nonconformity will require the preparation of a corrective action report (CAR). QVS Corp.'s CAR is designed to satisfy the following:
  a. Determine the causes of nonconformities or potential nonconformities
  b. Evaluate the need for action to ensure that nonconformities do not occur or reoccur
  c. Determine and implement the appropriate action needed
  d. Record the results of actions taken.
  e. Review the effectiveness of the action taken.

CARs will be tracked to ensure the correction has been made. When the nonconformity has been corrected, the CAR will be filed in the appropriate Microsoft SharePoint file.

Figure 6.13 QVS Corporation's nonconformities and corrective and preventive action process.

CONTROL OF RECORDS

ISO 50001 EnMS Standard

Purpose: To control records that meet the standard requirements and demonstrate the energy performance. The controls should include identification, retrieval, and retention of the records.

Control of records characteristics: The records should be legible, easy to identify and retrieve, and retired in accordance with the organization's retention policy.

Operational Explanation

The retention of records is four years. QVS Corporation's retention policy also applies to the EnMS records. Records will be kept in the appropriate file folders in the centralized EnMS document control system on Microsoft SharePoint. They should be legible, easy to retrieve, and controlled. The controls should include adding a title and the origination date to each record and adding a revision number when revised. Only people responsible for changing a record will be given access to make the change. Other interested personnel will have read-only access to the records. Once the documents become a record, they may no longer be changed since they are no longer active.

Forms, Templates, Processes, and Plans to Meet the Standard

Records are kept for four years in accordance with QVS Corporation's retention policy for records. Periodic checks, similar to those described for the CAR earlier, will help ensure documents are controlled.

QVS Corporation Example of Implementation

The energy team's document control manager looks at all documents once a year to see whether any are over four years old. The origination date on the document enables this review. If the document is a record that is more than four years old, it is pulled from the document control system and retired in accordance with the company's retention policy. If the document is more than four years old but is not a record (still useful to the team and may be used), it will remain in the document control system. It may be stamped "reviewed—still useful."

Chapter 7
Management Review

GENERAL REQUIREMENTS
ISO 50001 EnMS Standard

Purpose: Management should conduct management reviews at planned intervals such as once a year and review the EnMS for suitability, adequacy, and effectiveness.

Management review characteristics: Management reviews should be held at planned intervals (at least once a year), all participants should receive a notice and the agenda before the meeting, the appointed top manager(s) should attend, the energy team leader should be the master of ceremonies, all required inputs should be covered, and all required outputs should be produced. The documentation will include the agenda, the sign-in sheet, presentations given, and the minutes of the management review.

Operational Explanation

Management reviews are critical for continuous improvement. ISO standards require them. They are part of the "check" phase and a large part of the "act" phase. The standard says management reviews should be conducted at planned intervals, and most management reviews are conducted at least annually by the organization's senior management. In a management review, the management evaluates the progress of current projects and O&Ts and determines whether there are any barriers. They recommend new projects and objectives. They review any internal or external audits. Their objective is to improve system performance and ensure that the required inputs and outputs by paragraph 4.7 of the ISO 50001 EnMS are met. The management review gives management the opportunity to see whether the standard being implemented is suitable, acceptable, and effective for the organization.

Management reviews should be held at least once a year. Top management and the energy team should be invited. The energy champion or his or her boss chairs the meeting. The energy team leader or his or her designee serves as the master of ceremonies.

Forms, Templates, Processes, and Plans to Meet the Standard

The note taker should keep the minutes of the management review. The agenda, meeting minutes, PowerPoint or other presentation, and the sign-in sheet should all be maintained in the document control system.

QVS Corporation Example of Implementation

The company's management representative, the energy champion, leads the management review. It is conducted at least annually. The agenda includes all the input requirements, and the minutes capture all the outputs required.

INPUTS TO MANAGEMENT REVIEW

ISO 50001 EnMS Standard

Purpose: The inputs necessary to ensure management reviews accomplish their purpose of assessing suitability, adequacy, and effectiveness are outlined and should be included in every management review.

Management review inputs: The agenda should include the required inputs, and each should be discussed. First, the agenda should call for any follow-up actions from any previous reviews, followed by discussion on energy policy and how it is affected. The current status of the EnPIs should be discussed, along with the extent to which O&Ts are being met. If it is an EnMS audit or a self-inspection, the results should be outlined and discussed, including the status of any CARs or PARs and further actions needed. The energy performance status should be covered. Has it improved? If so, what has been achieved? Recommendations of improvements, including energy efficiencies, should be discussed and decisions made.

Operational Explanation

The energy champion, the sustainability executive officer, or a top management group such as the Strategic Council, Quality Council, Leadership Council, or other group conducts the management review to determine whether the EnMS being implemented is suitable, adequate, and effective. It is important in planning the meeting, and especially in preparing the agenda, that all required inputs are included and that the required outputs are produced and documented. The inputs should be:

1. Energy management action plan reviews, energy diagnoses/review results, and EnMS audit results (this includes changes to EnPIs)

2. Evaluation of legal and other compliance and any changes to legal or other requirements

3. The energy performance of the organization (how it is doing relative to EnPIs)

4. The status of corrective and preventive actions

5. The performance of the EnMS
6. The extent to which energy O&Ts have been met
7. Recommendations for improvement
8. Follow-up actions from previous management reviews

Any self-inspection or external energy audits need to be reviewed. Any CARs or nonconformities need to be discussed.

Forms, Templates, Processes, and Plans to Meet the Standard

1. Copies of self-inspections, energy audits, and CARs should be part of the management review.
2. Any business not resolved at the last meeting should be reviewed at this meeting.
3. Any action items or recommendations from the last meeting should be reviewed at this meeting.
4. Copies of current EnPIs should be reviewed. Any changes in targets, along with the results, should be shown at the management review.
5. The legal requirements should be reviewed at least annually as to the organization's compliance with them. It is a best practice to do that the month before the scheduled management review.

QVS Corporation Example of Implementation

The inputs required should be placed on the management review agenda and should be covered at the management review. The agenda from QVS Corporation's last management review is shown in Figure 7.1.

Time	Topic	Responsible
09:00–09:05	Introductions and management review requirements	Team leader
09:05–09:08	Legal requirements and other facts	Team leader
09:08–09:10	Self-inspection and corrective actions	Team leader
09:10–09:15	QVS energy policy	Team leader
09:15–09:20	EnPIs (energy performance indicators)	Team leader
09:20–09:25	EnPI baseline data and energy indicator performance	Team leader
09:25–09:30	Completed objectives and targets (O&Ts)	Team leader
09:30–09:40	Current O&Ts	Objective responsible persons
09:40–09:43	Completed projects and current projects	Team leader
09:43–09:45	Changes in energy policy and EnPIs	Team leader
09:45–09:50	Planned future actions	Team leader
09:50–10:00	Questions, answers, approvals, and recommendations	Objective champion

Figure 7.1 QVS Corporation management review agenda (January 15, 2013).

The agenda is included in the Microsoft PowerPoint presentation developed for this management review (see Appendix A). The agenda should show the date, the start and end times for each segment or item, all relevant inputs, and who is covering each item.

OUTPUTS FROM MANAGEMENT REVIEW

ISO 50001 EnMS Standard

Purpose: Management reviews are an essential part of the "check" and "act" phases of continually improving an organization's EnMS. The outputs required enable these necessary improvements and changes. As the management review is conducted, the outputs, which are now deliverables, are achieved as they are discussed, and decisions and recommendations are made.

Management review outputs: Any changes in the energy performance of the organization, in the energy policy, or in any of the EnPIs or O&Ts or any other element of the EnMS; any resource allocations; or any new projects should be covered in the meeting minutes and distributed to all interested parties.

Operational Explanation

Minutes of the management review must be taken, coordinated with all attendees, and approved by the energy champion or energy team leader. It is a best practice or a BAT to develop a Microsoft PowerPoint presentation for the meeting. In writing the minutes, the requirement can be copied from the Microsoft PowerPoint slide, and then the decisions or actions recommended can be covered after the requirement. The format would include the following categories:

- Management review—organization name
- Purpose
- Date/time
- Place
- Requirement/decisions/recommended actions/action items
- Summary of energy champion or top management approvals and recommendations
- Approval by management review leader

All action items should be numbered and show who is to do what, when, and where, if appropriate.

As should be done for all energy team meetings and quarterly reviews, have all participants sign a sign-in sheet and document it in the same folder (in the central document control system) with the management review agenda, the minutes, and the Microsoft PowerPoint presentation. This will make it easy for the self-inspection team, the internal auditor, or an external auditor to locate.

A typical sign-in sheet contains the following info:

- Name
- Organization title
- Telephone number
- E-mail address
- Signature

The top of the sign-in sheet would have the organization's name, the title "Management Review," and the date.

Forms, Templates, Processes, and Plans to Meet the Standard

Meeting minutes of the management review, including action items and recommendations, should be documented. At the first energy team meeting after the management review, the team should review the recommendations, decide how they are going to be implemented, and initiate the action needed. The status of action items should be reviewed at each energy team meeting until they are implemented or resolved. The meeting minutes, agenda, presentations, and sign-in sheet should all be included in the document control system and filed by year completed. A sign-in sheet should be used at each meeting, including management reviews, showing the date and type of meeting and the info shown in the list.

QVS Corporation Example of Implementation

The meeting minutes should reflect all outputs, action items, new O&Ts, and the top manager's or management's determination as to the suitability, feasibility, and adequacy of the EnMS presented. The minutes should be in the format shown in Figure 7.2. The action items are reviewed at each regular meeting until they are implemented or resolved. Any action items that are not completed prior to the next management review are placed on the new agenda and discussed. Notice how the Microsoft PowerPoint slide presentation for the management review in Appendix A is used to develop the minutes. It covers the information that was presented and discussed, what was approved, and the recommendations.

Date/time: January 15, 2013, 9:00–10:00 AM

Place: QVS Corp.'s Executive Conference Room

Purpose: To conduct a management review of QVS Corp.'s development and implementation of ISO 50001 EnMS.

Agenda:

Time	Topic	Responsible
09:00–09:05	Introductions and management review requirements	Team leader
09:05–09:08	Legal requirements and other facts	Team leader
09:08–09:10	Self-inspection and corrective actions	Team leader
09:10–09:15	QVS energy policy	Team leader
09:15–09:20	EnPIs (energy performance indicators)	Team leader
09:20–09:25	EnPI baseline data and energy indicator performance	Team leader
09:25–09:30	Completed objectives and targets (O&Ts)	Team leader
09:30–09:40	Current O&Ts	Objective responsible persons
09:40–09:43	Completed projects and current projects	Team leader
09:43–09:45	Changes in energy policy and EnPIs	Team leader
09:45–09:50	Planned future actions	Team leader
09:50–10:00	Questions, answers, approvals, and recommendations	Objective champion

Required inputs:

- Energy management action plan reviews, energy diagnoses/review results, EnMS audit results (this includes changes to EnPIs)
- Evaluation of legal and other compliance and any changes to legal requirements
- Evaluation of the energy performance of the organization (how are we doing relative to EnPIs?)
- Review of the status of corrective and preventive actions
- Evaluation of the performance of the EnMS
- Assessment of the extent to which energy O&Ts have been met
- Recommendations for improvement

Required outputs:

- The improvement in the energy performance of the organization since the last review
- Changes to the energy policy
- Decisions regarding the energy performance of the organization
- Decisions regarding the EnMS
- The validity/suitability of EnPIs
- Changes to the objectives, targets, or other elements of the EnMS consistent with the organization's commitment to continual improvement
- Allocation of resources

Legal requirements and other factors:

- Electricity provider is Gun Barrel City Electric Company. Does not provide a UESC (utility energy savings contract) service.
- Can pursue another electric provider since Texas is a deregulated state.
- Must keep power factor above 0.95 or pay an adjustment fee.
- Must keep all electric conduit and system inside the facilities maintained in workable condition.
- Present cost of electricity is 6.5 cents per kWh.

Figure 7.2 QVS Corporation management review minutes.

Corrective actions:
- No outside or second-party audits.
- In October 2012, the energy team performed a self-inspection. All discrepancies (primarily in documentation) have been corrected.
- Next self-inspection is scheduled for October 2013.

QVS Corp.'s energy policy:

QVS Corp. is committed to purchasing and using energy in the most efficient, cost-effective, and environmentally responsible manner possible. Therefore, QVS Corp. shall:
- Practice energy conservation at all facilities
- Lower peak demand at all facilities
- Improve energy efficiency while maintaining a safe and comfortable work environment
- Lower kWh per square foot to best-in-class levels
- Increase the percentage of renewable energy used
- Continually improve its performance

EnPIs:
- kWh usage by month and summed for year
- kWh usage/gross square footage summed for year
- 2010 is our baseline year

   kWh usage and baseline (2010):

Headquarters:	2,681,740
Plant A:	3,495,709
Plant B:	3,423,075
Plant C:	3,275,166
Plant D:	3,245,210
QVS total:	16,120,900

   Electricity intensity baseline (kWh/sq. ft.):

Headquarters:	48.76
Plant A:	34.10
Plant B:	27.38
Plant C:	33.42
Plant D:	32.45
QVS:	33.55

- QVS goal is to reduce kWh usage by 10% by end of 2015

   Actual vs. targets:

Results at end of year 2011:	0.16%
At end of year 2012:	3%
Target at end of year 2013:	5%
At end of year 2014:	7%
At end of year 2015:	10%

Completed O&Ts:
- OT-11-01 Energy Awareness Training: In 2011, all employees and contractors were trained. They will be retrained in 2014.
- OT-11-02 Energy Competency Training: Facility managers were also made the energy managers. An energy manager training course was developed and presented by the team leader and the facilitator that included their roles and responsibilities, basic terms, EnPIs, things to look for to achieve savings, the energy conservation program, and how they can help.

Figure 7.2 QVS Corporation management review minutes. *(Continued)* (continued)

- OT-11-03 Communication Plan: A communication plan has been developed to show how communications on energy should be conducted both in-house and outside the organization. All employees have been trained on the contents of the plan.
- OT-11-05 Energy Conservation Program: An energy conservation program was developed, put into Microsoft PowerPoint, and sent to all employees and contractors by the objective champion in late November 2011.

Current O&Ts:

- OT-11-04 Develop and Implement a Purchase Control Plan: Actions revised to include renegotiating with present provider, and others available due to deregulation, and to lower the present cost of 6.5 cents per kWh. In addition, QVS wishes to raise the percentage of renewable energy from present 2% to 15%.
- OT-12-01 Implement an Information Technology (IT) Power Management Program: The sleep function was enabled on all computer monitors, and the hibernate function was enabled on 25% of computers and 10% of laptops. New electronics purchased must be Energy Star rated, and electronics that have reached their end of life will be recycled with an R2 electronic recycler.

Completed projects:

HQ: Changed T12 to T5 lights (completed on November 30, 2011)
Plant A: Changed T12 to T5 lights (completed on December 10, 2011)
Plant B: Changed T12 to T5 lights (completed on December 30, 2011)
HQ, Plants A–D: Installed 120 occupancy sensors in areas of infrequent use (restrooms, break rooms, mechanical rooms, and copier rooms)
Plants A, C, and D: Installed 12 inches of insulation in the roof ceilings
Plant D: Replaced exit signs with low-energy-use exit signs

Future projects/ECMs:

- Replace cooling tower in HQ building. Payback period is 8.2 years.
- Replace boiler in Plant B. Payback period is 9.5 years.
- Replace hot water heaters in HQ building and Plants A, B, and D with solar panels. Payback period is 3.7 years.

Changes in QVS Corp.'s energy policy:

QVS Corp. is committed to purchasing and using energy in the most efficient, cost-effective, and environmentally responsible manner possible. Therefore, QVS Corp. shall:
- Practice energy conservation at all facilities
- <u>Lower peak demand at all facilities</u>
- Improve energy efficiency while maintaining a safe and comfortable work environment
- Lower kWh per square foot to best-in-class levels
- <u>Increase the percentage of renewable energy used</u>
- Continually improve its performance

(Underlined items in list above need to be addressed.)

Energy policy recommendations:

#1. Keep energy policy as is.
#2. Recommend a new O&T be approved to address peak load.
#3. Get Strategic Council to approve 15% by end of 2015 as our renewable energy goal or target.

The energy champion approved recommendations 2 and 3. **Action item #1:** The energy champion will present recommendation 3 to the Strategic Council for improvement at its next meeting on April 12, 2012. **Action item #2:** The energy team will develop an

Figure 7.2 QVS Corporation management review minutes. *(Continued)*

O&T and energy action plan at its next meeting in February 2012 and then implement the energy action plan.

#4: Keep EnPIs (kWh usage by month and summed for year, kWh usage/gross square footage summed for year). They are valid and suitable.

#5: For the O&T to reduce the peak load, recommend that we add electric load factor (ELF) to be calculated quarterly to ensure our peak loads have not changed negatively for some reason.

- ELF is an indicator that shows if peak demand is high for your facility. It is an indicator of how steady an electrical load is over time. The optimum load factor is 1 or 100%. The closer to zero, the more you are paying for electricity.

$$ELF\ (\%) = \text{Total kWh} / \text{\# days in electricity billing cycle} \times 24\ \text{hours/day} / \text{peak kW demand} \times 100$$

In other words, it is the average demand/peak demand for a given period of time. From your electricity bill get the kWhs used and the peak kW. Next, look for days included in the bill. Multiply these days by 24 (hours per day). Divide this number into the kWhs. Then divide what you get by the peak kW. Multiply this number by 100 to get a percentage.

The energy champion approved recommendation 4. He recommended that the ELF be the measure for the new O&T approved above. **Action item #3:** The energy team will include development and tracking of the ELF for each QVS Corp. facility at the next energy team meeting in February 2012.

Planned future actions:

- Continue meeting monthly even if only 30 minutes in duration.
- Manage O&Ts and projects/ECMs to completion.
- Standardize the objective champion's quarterly update to the Strategic Council.
- Continue to identify ECMs.
- Perform second self-inspection in October 2012.

The energy champion approved the future plans and complimented the team on its past actions and achievements. **Action item #4:** The energy team needs to finish its recommendations for standardizing the Strategic Council's quarterly update and present it to the energy champion by March 5, 2012. **Action item #5:** The energy team leader and the facilitator shall develop the agenda for the next energy team meeting in February 2012 by February 3, 2012, and communicate to all involved personnel. **Action item #6:** The energy team needs to plan for conducting a second self-inspection in October 2012 and sharing the results with the energy champion.

Recommendations and approvals by energy champion:

1. Approved current O&Ts.
2. Recommended new O&T for measuring and improving the ELF.
3. Approved keeping existing EnPIs.
4. Approved future planned actions for the energy team.

Approved:
Energy Champion

Figure 7.2 QVS Corporation management review minutes. *(Continued)*

Chapter 8
Integration of ISO Standards

INTEGRATION

Most companies considering using ISO 50001 EnMS have already implemented ISO 9001 QMS and possibly ISO 14001 EMS and OHSMS 18000 Safety. ISO 50001 EnMS used ISO 14001 EMS as a guide in its development. ISO 14001 EMS used ISO 9001 QMS as a guide in its development. Therefore, there are a lot of similarities in the phases and elements that are required. It is easy for a company to just add ISO 50001 EnMS to the systems it already has. How did QVS Corporation approach this possibility?

QVS CORPORATION INTEGRATION OF ISOs

After the EnMS was developed and had been in place for about a year, the Strategic Council asked the energy champion to explore the feasibility of merging ISO 9001 QMS, ISO 14001 EMS, and ISO 50001 EnMS. To do so, he felt a new integrated and cross-functional team would be necessary. He asked the different functional departments if the energy team leader, the environmental management team leader, and the quality management systems team leader could be part of the new integrated standards team (IST). The facilitator for the energy team was asked to facilitate the IST. The energy champion would serve as the team leader. A corporate lawyer and a member of the quality department rounded out the team. The purpose of the team's first meeting was to establish the ground rules and identify elements that could be consolidated. The team members reviewed the elements of each ISO standard prior to the meeting.

First Meeting

The IST adopted the energy team's ground rules since they fit the needs of the new team. The IST decided not to include ISO 9001 QMS in the integration. ISO 9001 QMS has existed for several years and the documentation is extensive. The quality manual has undergone several revisions, which have kept it current. ISO 9001 has received several successful audits from outside auditors. ISO 14001 and 50001 have a lot of elements in common. The team reviewed the standards and identified the following elements as ones that could be integrated or combined:

1. Roles, responsibilities, and actions
2. Policies
3. Legal and other requirements
4. O&Ts
5. Awareness training
6. Operational controls
7. Documentation requirements
8. Communications
9. Purchasing
10. Monitoring and measurement
11. Evaluation of legal and other requirements
12. Management system audits
13. Nonconformities and corrective, preventive, and improvement actions
14. Control of records
15. Management reviews

Second Meeting

The purpose of this meeting was to evaluate the elements that had been earmarked for possible integration (see Table 8.1). The identified actions are now the IST's action plan.

Third Meeting

The purpose of this meeting was to review both the environmental policy and the energy policy and, taking the salient points, write a compelling environmental and energy policy. Figure 8.1 shows QVS Corporation's environmental and energy policy.

Fourth Meeting

The purpose of this meeting was to review the roles and responsibilities for both the EMS and the EnMS and consolidate them. The team defined the roles and responsibilities as shown in Figure 8.2.

Table 8.1 Evaluation of elements for possible integration.

Element	Fully combined?	Partially combined?	Actions needed	Status
1. Roles, responsibilities, and actions	Yes		Rewrite roles, responsibilities, and authority and make them for both ISO systems	
2. Policies	Yes		Combine into an environmental and energy policy	
3. Legal and other requirements		No—legal requirements are different for each	Place in "Legal Requirements" file but in separate folders in Microsoft SharePoint	
4. O&Ts	Yes		The O&T and action plan templates can be used for both; need to convert the EMS O&Ts and action plans to the energy template	
5. Awareness training	Yes		Need to combine the awareness training for both environmental and energy policies and send to all employees and contractors	
6. Operational controls		No—operational controls are different and combination would be confusing	Place in "Operational Controls" file but in different folders in Microsoft SharePoint	
7. Documentation requirements	Yes		Place in same file with the element name but in different folders in Microsoft SharePoint	
8. Communications	Yes		Consolidate both the energy and environmental communications plans into one and make visible to all personnel	

(continued)

Table 8.1 Evaluation of elements for possible integration. *(Continued)*

Element	Fully combined?	Partially combined?	Actions needed	Status
9. Purchasing	Yes—each purchasing requirement is different except for power management		Combine into one purchasing plan	
10. Monitoring and measurement	Yes		Combine into one monitoring and measurement plan	
11. Evaluation of legal and other requirements		No—legal requirements are much different	Place in "Evaluation of Legal Requirements" file but in different folders in Microsoft SharePoint	
12. Management system audits		No—self-inspections and audits should be conducted separately	Place in "Audits" file but in separate folders in Microsoft SharePoint	
13. Nonconformities and corrective, preventive, and improvement actions		No	Place in "Corrective Actions" file but in separate folders in Microsoft SharePoint; use the CAR and PAR process for both systems	
14. Control of records	Yes		Place both in same file in Microsoft SharePoint but in different folders	
15. Management reviews	Yes		Can conduct both ISO systems' management reviews at same time, will need to add all the top management reviewers who participated in both systems' past management reviews	

In compliance with the environmental and energy laws of the United States and the state of Texas, QVS Corp. is fully committed to performing its activities in a manner that demonstrates leadership in environmental and energy stewardship and sustainment. QVS Corp. realizes that preserving the nation's natural resources and protecting the environment are objectives that are aligned with corporate long-range goals of serving its customers on time with quality products. QVS Corp. is committed to reducing negative impacts on the environment that result from producing its products and consuming energy.

QVS Corp. is committed to complying with both ISO 14001 Environmental Management System (EMS) and ISO 50001 Energy Management System (EnMS). QVS Corp. will strive to achieve these standards through sustained environmental stewardship, recycling, pollution prevention efforts, energy conservation, energy reduction projects, and reduction of greenhouse gas (GHG) emissions. QVS Corp. will implement, maintain, and sustain EMS teams at the headquarters, plants, and distribution center and use them as the primary mechanism to achieve and sustain environmental excellence. A headquarters energy team with points of contact (POCs) has been established for energy planning and reduction. Both the EMS champion and the energy champion will report to QVS Corp.'s Strategic Council. Together these two champions will manage these strategic stewardship efforts and achieve the environmental and energy goals of QVS Corp.

All QVS Corp.'s managers, employees, and contractors must incorporate environmental and energy stewardship into their decision making and day-to-day activities to protect the public and land, water, air, natural resources, and energy sources. QVS Corp. will integrate environmental stewardship and energy reduction sustainability principles into its core mission to the extent feasible by incorporating the following objectives:

- Improve the energy efficiency of buildings, equipment, vehicles, travel, and other operational factors in order to reduce GHGs.
- Promote resource conservation by managing water usage, reducing energy usage, and implementing renewable energy options.
- Integrate environmental impact considerations and environmental and energy management principles into planning, purchasing, operating, and budgeting decisions.
- Plan, build, and operate highly sustainable new buildings and improve existing buildings' environmental and energy efficiency.
- Implement pollution prevention practices through sustainable acquisition, electronic stewardship, recycling, and other waste diversion efforts.
- Purchase environmental-preferred products whenever possible.
- Comply with national, state, and local environment and energy laws, statutes, and building codes.
- Continue environmental and energy performance through EMS and EnMS at its facilities and monitor the results.
- Communicate all results to employees and contractors and the public.

QVS Corp. is dedicated to becoming a leader in both environmental and energy management leadership and will always strive to save water, reduce environmental risk at its facilities, reduce energy consumption, and increase its use of renewable energy.

John D. Money, CEO, QVS Corp.

Figure 8.1 QVS Corporation environmental and energy stewardship policy.

Head of organization
1. Be responsibile for the overall development, implementation, and maintenance of the EMS and EnMS
2. Review and approve the essential documents written and issued and decisions made by the EMS (environmental teams) and EnMS (energy team)
3. Provide adequate resources for the EMS and EnMS implementation

Instead of the head of the organization, this could be the facility head. The head of QVS Corporation delegated this responsibility to the Strategic Council, which in turn appointed a management representative for the EnMS (the energy champion) and a management representative for the EMS (the environmental champion). The environmental champion is the functional director of environmental health and safety.

Team leader
1. Work with the facilitator in developing the meeting agendas
2. Lead each team meeting
3. Ensure the requirements of the ISO 50001 EnMS and ISO 14001 EMS standards are met
4. Provide quarterly updates on progress and results to the Strategic Council
5. Ensure O&Ts are developed and implemented
6. Lead the development of the management review inputs and outputs and be the master of ceremonies at the management review
7. Approve the team meeting minutes
8. Represent the team in other meetings and activities where required

Some teams find it helpful to have an assistant team leader if the team leader has to travel a lot. Often an operations manager is appointed to ensure objectives, targets, and monthly milestones are achieved. This individual can also serve as an assistant team leader.

Note taker
1. Take notes during the meeting
2. Serve as a scribe if needed by the team leader
3. Write up the meeting minutes and distribute them in a timely manner to the meeting participants and interested others
4. Ensure that the minutes are filed in the correct file and folder on Microsoft SharePoint

Document control manager
1. Maintain documents and records so they are easy to retrieve
2. Ensure that the team uses the most current documents
3. Ensure that the established filing system on Microsoft SharePoint is properly labeled and that documents or records are in the correct file or folder

Often the team leader volunteers to be the document control manager since he or she is involved with every document produced by the energy team. The environmental teams produce more documents, so the document control managers for the energy team and those for the environmental teams, along with their respective facilitators, were retained as document control managers. They, along with the team leaders, have the authority to place documents and delete documents or records. The others only have read-only authorization.

Team members
1. Assist the team leader in achieving meeting purposes
2. Identify elements that can be integrated
3. Assist in achieving O&Ts
4. Identify possible improvements, including new objectives and projects
5. Communicate information about the energy program to the appropriate facility staff

Figure 8.2 QVS Corporation's roles and responsibilities during team meetings.

Facilitator

Before the meeting:

1. Make sure arrangements have been made for a meeting place
2. Prepare an that gives the meeting's purpose, location, date and starting time, and items with a time frame for discussion and the responsible person or persons for each, and send to all participants prior to the meeting
3. Be sure handouts, critique forms, Microsoft PowerPoint slides, and sign-in sheets are available

During the meeting:

1. Build teamwork
2. Manage conflict
3. Keep team task and working toward the meeting's purpose
4. Achieve participation
5. Help team reach consensus when possible
6. Evaluate and critique the meeting's effectiveness
7. Ensure that the right tool or technique is used correctly
8. Highlight action items and decisions and ensure that the team meets the needs of a decision and its effects before moving forward

After the meeting:

1. Assist any team member in accomplishing his/her task if needed
2. Ensure meeting minutes are accurate, complete, and distributed in a timely manner
3. Assist team leader in briefing management and employees on the status, barriers, issues, and accomplishments

Figure 8.2 QVS Corporation's roles and responsibilities during team meetings. *(Continued)*

Fifth Meeting

The purpose of this meeting was to combine the EMS and EnMS files on Microsoft SharePoint into one set of files. The new files were titled "Environmental and Energy Management Systems." The files consist of the documentation in Table 8.2.

Sixth Meeting

The purpose of this meeting was to consolidate the coming year's annual training into a program that has both environmental and energy training.

Seventh Meeting

The purpose of this meeting was to consolidate the company's communication plan into one that includes both environmental and energy awareness (Figure 8.3).

Table 8.2 Integrated documentation.

Title	EMS	EnMS	Additional instructions	Additional requirements
1. Aspects	Aspects, impacts, and risk analysis	Significant electricity users, significant natural gas users		Have folder for each year
2. Legal requirements	National, state, city, and community laws, statutes, and other requirements	National, state, and city laws, statutes, and other requirements, and building codes		Have folder for each year
3. O&Ts and action plans	Current O&Ts and action plans for environmental	Current O&Ts and action plans for energy	File by year and place others in a folder by year of origination	O&Ts and action plans are numbered by year. In the central documentation file, keep all O&Ts that are still being worked on in the current file. Once completed, place in the year of origination file.
4. Projects	Any environmental projects that use contractors to complete	Any energy projects that were advertised and a contract awarded		Have folder for each year
5. Training	Any specific environmental training and sign-in sheets	Any specific energy training and sign-in sheets	Place combined environmental and energy training in a combined awareness training folder	Have folder for each year
6. Monitoring and measurement	Place the energy monitoring and measurement list and all EnPIs in the central documentation system on Microsoft SharePoint	Place the environmental monitoring and measurement list and all EnPIs in the central documentation system on Microsoft SharePoint		Have folder for each year

		Self-inspections done for EMS	Self-inspections done for EnMS	
7.	Self-inspections	Self-inspections done for EMS	Self-inspections done for EnMS	Have folder for each year
8.	Nonconformities	CARs/PARS	CARs/PARs	Use same form
9.	Communications	Place internal and external environmental communication (internal includes meeting agendas, minutes, and sign-in sheets)	Place internal and external energy communication (internal includes meeting agendas, minutes, and sign-in sheets)	Put the combined communications plan in the combined communications folder
10.	Operational controls	Place current operational controls for environment	Place current operational controls for energy	Have folder for each year
11.	Management reviews	Management reviews for environmental (presentations, agenda, minutes, and sign-in sheets)	Management reviews for energy (presentations, agenda, minutes, and sign-in sheets)	Consider combining in 2014
12.	Annual reviews	Every November, show what has been accomplished during year	Every November, show what has been accomplished during year	Consider combining in 2014
13.	Plans	Emergency preparation plan	Contingency plan	Have folder for each year

---

Memorandum

Subject: QVS Corp. Energy Communications Plan   Date: January 27, 2011
To: All QVS Corp. personnel
From: Gene Smith, President, QVS Corp.

Purpose
This memorandum describes QVS Corporation's policy for both internal and external communications in regard to QVS Corp.'s Energy Management System (EnMS) and Environmental Management System (EMS). The specific purpose is to spell out the methods and processes for communicating with QVS Corp.'s management, employees, and contractors, and instruct how to handle external inquiries for information.

Requirement/Scope
This memorandum is in response to paragraph 4.4.3 Communications, ISO 14001 EMS and 4.5.3 Communications, ISO 50001 EnMS.

Definitions
Communications media: A medium is one specific method of communicating. E-mail is a medium. A letter is a medium. An announcement in a staff meeting is a medium. Other media include signage, pamphlets, handouts, in-person training, loudspeaker announcements, intranet, notice boards, bulletin boards, kaizen events, town hall meetings, seminars, training sessions, and telephone calls.

EnMS: The Energy Management System in accordance with ISO 50001 Energy Management System (EnMS).

Major environmental concerns: Concerns that deal with legal requirements not being implemented or followed that could lead to imminent human health concerns, or an imminent adverse environmental impact, fines, or legal action being taken against a facility or its employees. Major environmental concerns can be considered emergency communications.

Major energy concerns: Concerns that deal with legal requirements not being implemented or followed that could lead to adverse situations, for example, the potential loss of power for prolonged periods due to weather or legal concerns or a contingency plan becoming inoperable for some unforeseen conditions. Major energy concerns can be considered emergency communications.

Minor environmental and energy concerns: Concerns that deal with environmental or energy issues that can be addressed or improved, but have no imminent health, environmental, or legal repercussions. Minor environmental and energy concerns can be dealt with through a facility's EMS and energy team as necessary. Examples are any environmental, health, or safety information that does not require immediate action to solve or disseminate, such as an incorrect label on a container, proper inventory levels, a safety rule interpretation, what shots will be necessary to travel next month to an overseas location, and so forth. The answer can be obtained during the next three days and the issue does not in any way become an emergency situation or problem.

Formal environmental requirement: A letter or e-mail request for information in regard to a procedure, audit, or similar request that spells out what is desired and by when.

Informal environmental requirement: A telephone or in-person request for specific information that may or may not have an associated timetable. The request could come from a QVS Corp. person or an external person.

Formal energy requirement: A letter or e-mail request for information in regard to a procedure, audit, or similar request that spells out what is desired and by when.

Figure 8.3 QVS Corporation's integrated communication plan.

Informal energy requirement: A telephone or in-person request for specific information that may or may not have an associated timetable. The request could come from a QVS Corp. person or an external person.

POC: Primary point of contact.

Staff: The people who work at the facility, including management and employees.

Responsibilities

The environmental champion and the energy champion or their designees are responsible for documenting and keeping records of internal and external major environmental or energy concerns that reach their level for action or forwarding to the Strategic Council. The environmental champion and the energy champion will be informed of all major environmental concerns and energy concerns, respectively, and all minor concerns.

The EMS teams and the EnMS or energy team are responsible for keeping QVS Corp.'s personnel and contractors informed on EMS- and EnMS-related matters, to include how they can help improve its environmental or energy performance. The EMS and EnMS teams will keep records of all internal or external, major or minor environmental or energy concern communications that they deal with, including any responses. The documentation will be in the share drive integrated EMS and EnMS file in the communications folder under internal or external communications as appropriate. The files on the share drives are available to the staff for information as read-only. These files must be kept current by the EMS team document control managers and the EnMS document control manager. Letters, standard operating procedures (SOPs), work instructions, and company procedures are other communication tools available.

The environmental and energy champions are responsible for providing a quarterly update to the Strategic Council on EMS and EnMS progress, barriers, and performance. The EMS teams and the energy team are responsible for assisting their champions in preparing the quarterly updates.

The corporate energy team is responsible for providing annual energy awareness training and any competency training needed to ensure process and objective performance. The EMS team leaders and facilitators will assist the energy team in consolidating the energy and environmental awareness material into one training package that will be sent to all personnel by the energy champion.

The building's facility manager is responsible for handling all energy matters and answering energy questions for his/her facility. In the event of a major energy concern, the facility manager will notify the corporate energy team leader immediately.

Internal Communications Processes

Sending communications: The EMS teams and the EnMS or energy team will communicate with the staff using any appropriate media. To ensure that the message reaches everyone, using more than one medium is often required.

Receiving communications:

<u>Minor environmental and energy concerns:</u> Any minor energy concerns, suggestions, ideas, problems or perceived problems, or other EnMS-related communications should be reported to the energy team leader or any EnMS team member. Concerns relating to environmental or occupational health and safety aspects should be given to the environmental health and safety officer, who will either address them or elevate them to the level where they can be solved.

<u>Major environmental and energy concerns:</u> Major energy concerns should be immediately directed to a supervisor, manager, vice president, or a Strategic Council member. As appropriate, they will respond to communications. Any concern that cannot be addressed at the facility level will be directed to the corporate energy team leader or to the energy reduction objective champion. Major environmental concerns should first be sent to the facilities' environmental

Figure 8.3 QVS Corporation's integrated communication plan. *(Continued)*   *(continued)*

health and safety managers, but if they cannot be resolved at this level, the managers should send them to the environmental champion with recommendations on how they should be handled.

External Communications Processes
Receiving communications: Any QVS Corp. employee could receive an external request for information by receiving an e-mail, a letter, or a telephone call, or by having a visitor. This request should be made known immediately to a supervisor or corporate energy team leader, if energy related, who in turn will bring it to the energy reduction objective champion's attention (all external minor or major environmental concern requests). The energy champion will respond or send to the appropriate person or office for response. The request for environmental information should be given to the environmental health and safety manager, who in turn will coordinate with the corporate environmental champion to determine the appropriate course of action.

Both the environmental and the energy champions will determine which media inquiries should be referred to the Freedom of Information and Records Management Section at QVS Corp.'s headquarters.

All external inquiries with energy issues by members of the media or the public must be forwarded to the appropriate public information officer through the chain of command. Legal communications concerning the energy performance of the facility will be handled by the environmental or energy champion.

Points of Contact
The energy team leader is always a point of contact for minor energy concerns. In the event of a major environmental or energy concern, the employees should immediately report to their supervisor, who in turn will resolve or present to the environmental or energy champion for resolution. The EMS and EnMS team leaders are the primary points of contact for minor environmental and energy concerns.

Communication is the key to achieving our mission and meeting our EMS and EnMS responsibilities. Your past communications, job achievement, and conscientious attitude are most appreciated and needed in achieving our future responsibilities and mission. Protecting the environment and reducing energy usage are the responsibilities of everyone.

Figure 8.3 QVS Corporation's integrated communication plan. *(Continued)*

Eighth Meeting

The purpose of this meeting was to determine what changes to the O&T template are needed in order to use it for both energy and environmental teams. The O&T and action plan template is shown in Figure 8.4. To use as an environmental O&T, replace "Electricity high users addressed" with "Environmental aspect addressed" and replace "Energy action plan" with "Environmental action plan."

Ninth Meeting

The purpose of this meeting was to revise the present energy CAR so that it can be used for either energy or environmental nonconformities (see Figure 8.5).

Tenth Meeting

The purpose of this meeting was to develop a consolidated measuring and monitoring list (see Table 8.3).

Facility name: QVS Corporation, Gun Barrel City, Texas			Objective #: OT-11-001		
Objective: To develop a an electricity conservation program and kWh awareness training program.					
Target: Employees and contractors at QVS Corporation					
Initiation date: 1/2/2011		Anticipated completion date: 12/31/2015		Actual completion date:	
Electricity high users addressed: AC and heating, lighting, office equipment including computers and monitors, and other					
Baseline: 2011 electricity costs and kWh used by all facilities			Monitored or measured:		

Energy Action Plan					
#	Required action	Person responsible	Target date	Status	Comment
	List each step needed to ensure O&T is met.	Enter a name.	Enter the date the team expects this step to be done.	Enter "red," "yellow," or "green."	Enter the status of this step and record the date beside it (e.g., "Completed [4/4/11]" or "Management has not yet responded, extending target date by 10 days to 4/18/11 [4/4/11]").
1	Identify actions that can be taken to save energy.	EnMS team	2/3/2011	Green	
2	Select actions for electricity conservation program.	EnMS team	2/16/2011	Green	
3	Get actions approved by objective champion.	Team leader	2/20/2011	Green	Objective champion briefed Strategic Council and it liked the program.
4	Put actions into a PowerPoint presentation.	Bill Gibson	2/22/2011	Green	
5	Send to all employees and contractors.	Objective champion	3/2/2011	Green	

Figure 8.4 QVS Corporation O&T action plan.

Corrective Action Request			
CAR # (start with E or En):	Location:	Policy reference:	Date issued:
Requirement:			
Nonconformance:			
Corrective action issued by:		Corrective action assigned to:	
		Corrective action due by:	
Cause analysis EMS or EnMS team meeting date (if applicable): Root cause analysis:			
Corrective action Response: Supporting documentation:			
Acceptance Additional audits required? ☐ Yes ☐ No Corrective action accepted? ☐ Yes ☐ No Associated preventive action request (PAR) if applicable EMS/EnMS team leader _____ Date accepted _____ Office head _____ Date accepted _____			

Figure 8.5 QVS Corporation integrated CAR.

Instructions for preparing a CAR

Block 1 of form: Description of nonconforming work or departure from policies and procedures in the Environmental Management System (EMS) and/or Energy Management System (EnMS) or environmental or electricity regulations. This section records the following information:

a. CAR #: The name of the facility followed by "ECAR" or "EnMS CAR," the year, and a sequential numbering system (e.g., Plant A ECAR 12-001).
b. Policy reference: Enter the appropriate reference to a policy, procedure, or clause addressing the nonconformance.
c. Date issued: Enter the date that the CAR was issued.
d. Requirement: Enter a brief description of the requirement cited in the policy reference.
e. Nonconformance: Enter a detailed description of the nonconformance.
f. Corrective action issued by: Enter the name and title of the person issuing the CAR.
g. Corrective action assigned to: Enter the name of the individual assigned the corrective action.
h. Corrective action due by: Enter the deadline for the corrective action.

Block 2 of form: Cause analysis is the most important and sometimes the most difficult part of the corrective action process. The procedure for corrective action shall start with an investigation to determine the root cause(s) of the nonconformance. If necessary, the EMS or EnMS team should convene, discuss the nonconformance, and determine the root cause of the nonconformance. Potential root causes could include policies, methods and procedures, staff skills and training, and equipment. This section shall record a detailed description of the root cause analysis, while including any appropriate references and attachments.

Block 3 of form: Corrective actions are the steps implemented by the respective office to correct the nonconformance and prevent its recurrence. Corrective actions shall be appropriate to the magnitude and the risk of the nonconformance. This section shall record a detailed description of the corrective action chosen by the respective office. In addition, this section shall include all supporting documentation.

Block 4 of form: All documentation will be provided to the EMS or EnMS team leader for review. The EMS or EnMS team leader will determine whether further action is required. If no further action is required, the EMS or EnMS team leader will coordinate with the objective champion for concurrence. This section will be completed as follows:

a. Additional audits required: Check either "yes" or "no" to indicate whether additional audits are required. If "yes," provide reference to additional audit documentation.
b. Corrective action accepted: Check either "yes" or "no" to indicate whether the EMS or EnMS team leader and the objective champion have accepted the corrective action. If no corrective action is necessary, enter "yes" and indicate "N/A" with a brief description under the corrective action section.
c. Associated PAR: Enter the PAR number associated with the CAR (if applicable). The PAR can use this form by just writing PAR over the CAR heading.
d. Signatures and dates: Pending complete concurrence, the EMS or EnMS team leader and the objective champion will sign and date the CAR. By signing, both the EMS or EnMS team leader and the objective champion agree with all determinations (i.e., cause analysis, corrective action, etc.) and render the CAR complete.

Figure 8.5 QVS Corporation integrated CAR. *(Continued)*

Table 8.3 QVS Corporation's environmental and energy measurement and monitoring.

Aspect/item	Frequency	File location	Responsible person
Recycling	Monthly	Admin	Tom D.
Percentage of aspects recycled	Annually	EMS file	Facilitator
Electricity intensity	Monthly	EnMS file	Facilitator
Water intensity	Monthly	EMS file	Facilitator
Natural gas intensity	Monthly	EnMS file	Facilitator
Significant aspects	As needed (normally monthly)	AIRS and OTTS	EMS team
Power factor	Monthly	Facilities manager	Facilities manager
Operational controls	Annually	AIRS	EMS team
Number of items to landfill	Annually	EMS file	Facilitator
Hazardous waste generated	Monthly	EH&S manager	EH&S manager
Storm water retention	Monthly (visual)	EH&S manager	EH&S manager
Electric bill cost and kWh	Monthly	Facilities	Facilities manager
Water bill cost and gallons	Monthly	Facilities	Facilities manager
Natural gas cost and cubic feet	Monthly	Facilities	Facilities manager
Water permit	Semiannually	EH&S manager	EH&S manager
AIRS review	Annually or as needed	AIRS	EMS team
Current environmental and energy O&Ts	Monthly	OTTS	Operations managers

AIRS = Aspects, impacts, and risk spreadsheet
EH&S = Environmental health and safety
OTTS = Objectives and targets tracking sheet

Eleventh Meeting

The purpose of this meeting was to review the legal and other requirements and the operational controls for the EMS and EnMS to see whether they could be consolidated. The IST decided that, because they are so different, they should be filed in separated folders in the "Legal and Other Requirements and Operational Controls" file.

Twelfth Meeting

The purpose of this meeting was to determine whether the management reviews could be consolidated and, if so, what should be the inputs and outputs (see Figure 8.6).

Thirteenth Meeting

The purpose of this meeting was to review the self-inspection checklists to see whether they could be integrated. The IST felt it made more sense to keep the checklists and audits separate since they require different auditor expertise to conduct. The IST felt they should be conducted in the same month each year so that they can be presented at the management review.

Fourteenth Meeting

The purpose of this meeting was to consider combining the emergency preparation plan and the contingency plan. The team decided that this was feasible. See Figure 8.7 and Table 8.4 for the consolidated emergency and energy contingency plan.

Date/time: February 22, 2013, 10:00–11:00 AM

Place: Strategic Planning Conference Room

Purpose: To conduct an integrated management review for the EMS teams and the corporate energy team in accordance with ISO 14001 and 50001.

EMS purpose:

- Results of internal audits and evaluations of compliance
- Communications from interested parties, including complaints (if any)
- The environmental performance of the organization
- The extent to which O&Ts have been met
- Status of corrective and preventive actions (if any)
- Follow-up actions from previous management reviews
- Changing circumstances, including developments in legal and other requirements related to the facility's environmental aspects
- Recommendations for improvement, including approval of any new objectives and targets
- Feedback, input, and recommendations

EnMS inputs:

- Energy management action plan reviews, energy diagnoses/review results, EnMS audit results (this includes changes to EnPIs)
- Evaluation of legal and other compliance and any changes to legal requirements
- The energy performance of the organization (how it is doing relative to EnPIs)
- The status of corrective and preventive actions
- The performance of the management system for energy
- The extent to which energy O&Ts have been met
- Recommendations for improvement

Figure 8.6 QVS Corporation's integrated management review (EMS and EnMS) agenda. *(continued)*

> *EnMS outputs:*
> - Improvement in the energy performance of the organization since the last review
> - Changes to the energy policy
> - Decisions regarding the energy performance of the organization
> - Decisions regarding the EnMS
> - The validity/suitability of EnPIs
> - Changes to the O&Ts or other elements of the management system for energy consistent with the organization's commitment to continual improvement
> - Allocation of resources
>
> *Agenda:*
> - Results from internal audits/self-inspections (EMS and EnMS) and corrective actions
> - Legal and other requirements (EMS and EnMS)
> - Integration of EMS and EnMS status
> - Follow-up on actions from last management review
> - Completed O&Ts (EMS and EnMS)
> - Current O&Ts and responsible persons (EMS and EnMS)
> - Compliance
> - The environmental performance and energy performance of QVS Corp., including EMS indicators and EnPIs
> - Planned future actions
> - Recommendations for approval

Figure 8.6 QVS Corporation's integrated management review (EMS and EnMS) agenda. *(Continued)*

> *Requirements*
>
> ISO 50001 EnMS states, "The organization shall establish, document and maintain a procedure for identifying and responding to any energy supply or other potential disasters.
>
> This procedure shall seek to prevent or mitigate the consequences of any such occurrence and consider the continuity of the business operations."
>
> *Specific Actions Taken to Mitigate Loss of Power*
>
> In 2010 QVS Corp. experienced numerous service interruptions due to weather and to trees and squirrels interfering with power lines. Also, the power quality was not to the company's satisfaction since it negatively affected the motors and machines in the plants. QVS Corp. convinced the Gun Barrel City Electric Company to run a designated feeder (QVS Corp. was the only company on the feeder) from the new substation to the QVS Corp. complex. The old feeder remained. It was hooked up to another substation fed by another power plant. This arrangement significantly reduced the probability of QVS Corp. facilities losing power.

Figure 8.7 QVS Corporation's energy contingency plan.

Fifteenth Meeting

The purpose of this meeting was for the IST members to discuss whether they had missed anything (see Table 8.5) and to decide what should be presented to the Strategic Council for approval. One of the recommendations was to meet twice a year to consider further integration and replications.

Table 8.4 QVS Corporation's emergency preparedness plan.

Aspect	Activity, product, or service	Aspect name (impact)*	Emergency plan	Master plan location	Person responsible
16	Building-wide disposal of wastewater to sanitary sewer	Wastewater (HH, HW, RC/R, WP, WS/R, WU)	Hazardous communication plan	On Microsoft SharePoint, safety office	Environmental health and safety (EH&S) manager, safety officer
19	Waste disposal	Hazardous waste (AP, FH, HH, HW, GE, LF)	Environmental laws/regulations, occupant emergency plan (OEP), materials safety data sheets (MSDSs), hazardous communication plan	EH&S manager (laws/regs.), OEP (on Microsoft SharePoint), tool room (MSDSs), safety officer (hazardous communication plan)	EH&S manager
29	Equipment refueling, defueling aircraft and ground support equipment (GSE)	Fuel spill (AP, FH, GE, GC, HH, UP, RC/R, SC, SWC, WP, WS/R)	OEP, MSDSs, hazardous communication plan	OEP (on Microsoft SharePoint), tool room (MSDSs), safety officer (hazardous communication plan)	EH&S manager
41	Nickel cadmium and lead acid maintenance	Spills, fires, toxic gases, corrosives, heavy metals, absorbents (AP, FH, GE, GC, HH, HW, LF, UP, RC/NR, SC, SWC, UW, WP, WS/R, WS/NR, WU)	OEP	Security, on share drive	EH&S manager, security manager
60	Plants (nondestructive)	Corrosives (AP, FH, GE, HH, HW, LF, RC/NR, SC, SWC, UP, WP, WS/R)	OEP, MSDSs, hazardous communication plan	On Microsoft SharePoint, MSDSs located on plant floor in cabinets	EH&S manager
200	Supply receiving, inspection, shipping; tool room	Corrosives (FH, HH, HW, LF)	OEP, MSDSs, hazardous communication plan	On Microsoft SharePoint	EH&S manager

Environmental impacts key: AP, air pollution; FH, fire hazard; GC, ground contamination; GE, ground erosion; HH, health hazard; HW, hazardous waste; LF, land field; SC, soil contamination; SWC, storm water contamination; RC/R, resource consumption/recyclable; RC/NR, resource consumption/nonrecyclable; UW, universal waste; WP, water pollution

Table 8.5 IST meeting summary.

Meeting number	Meeting date	Purpose	Start/finish on time? (±5 min.)	Meeting facilitated?	Purpose achieved?	Participants (out of 10)	Deliverables completed?
1	1/10/2012	Establish ground rules, decide which ISO standards to integrate, identify elements to possibly be integrated	Yes	Yes	Yes	6	Ground rules, ISO 14001 EMS and ISO 50001 EnMS to be integrated, list of possible elements that can be integrated
2	1/25/2012	Evaluate the list of possible elements to be integrated, develop an action plan	No (+20 min.)	Yes	Yes	7	Elements to be integrated, action plan to integrate these elements
3	2/10/2012	Combine environmental policy and energy policy	Yes	Yes	Yes	7	An integrated environmental and energy policy
4	3/1/2012	Define joint roles and responsibilities for EMS and EnMS	Yes	Yes	Yes	5	Joint roles and responsibilities established
5	3/23/2012	Combine files on Microsoft SharePoint	Yes	Yes	Yes	6	Established integrated files on Microsoft SharePoint
6	4/14/2012	Consolidate training for both EMS and EnMS	Yes	Yes	Yes	5	Integrated awareness training
7	4/29/2012	Develop an integrated communications plan	Yes	Yes	Yes	6	Integrated communications plan
8	5/12/2012	Revise O&T template	Yes	Yes	Yes	6	Revised O&T and action plan template

		Revised CAR				Revised CAR
9	6/5/2012	Revised CAR	Yes	Yes	Yes	5
10	7/6/2012	Integrate the measuring and monitoring list	Yes	Yes	Yes	6	Integrated measuring and monitoring list
11	8/14/2012	Review legal and other requirements to determine feasibility of integrating	Yes	Yes	Yes	6	Decided not to have an integrated legal requirements list
12	9/12/2012	Consolidate management reviews	Yes	Yes	Yes	6	Consolidated management reviews, with inputs and outputs
13	10/12/2012	Review self-inspection checklists for possible integration	Yes	Yes	Yes	6	Decided not to combine
14	11/15/2012	Consider emergency preparedness plan and energy contingency plan for consolidation	Yes	Yes	Yes	6	An integrated emergency and contingency plan
15	12/10/2012	Review what has been done and determine whether anything else needs to be done	Yes	Yes	Yes	6	Everything is completed

The meeting metrics are satisfactory but not exceptional due to low member participation.
  Meetings facilitated: 100%
  Meetings achieving their purpose: 100%
  Meeting participation: 60%
  Meetings started and finished on time: 93%

ISO 14001 EMS and ISO 50001 EnMS were consolidated. This action saved time and led to other savings while not lessening either standard's maintenance efforts or results.

Chapter 9
Pitfalls and Countermeasures

MURPHY'S LAW

Murphy's Law states that what can go wrong will go wrong. In implementing the EnMS, the organization will likely observe Murphy's Law once or twice in the planning, developing, and implementing phases and/or during the internal audit. The hardest elements to accomplish are the identification of the SEUs and the measurement of all the items for which you would like to validate results. The identification of SEUs and their measurements should be considered a continuous process. Trying to do everything up front during the development and implementation phases may bog down your organization and lead to meaningful O&Ts being established late in the process. This is only one of the many pitfalls to avoid.

POSSIBLE PITFALLS AND COUNTERMEASURES

A pitfall is like a trap. You can fall into the trap thinking you are doing what is right but then not be able to accomplish your mission. When you experience a pitfall, you are following a path that looks correct but you discover later that you have veered off the correct course of action. Pitfalls can come from a variety of sources, such as no management support, no goals, insufficient resources, nonbelievers, and numerous others, as shown in the following list:

Pitfall 1: Not obtaining top management support prior to implementing ISO 50001 EnMS. Without top management involvement and support, the EnMS will eventually falter without any, or few, positive results. It is best not to advance too far into the EnMS process until management support is secured.

Countermeasure 1: Normally, this is not a problem since management initiates the call to implement the standard. If management does not do this, an energy team should develop an EnMS presentation that covers why the organization should implement ISO 50001 EnMS. The energy team should obtain functional approval to present to higher management. If higher management approves and supports the team with resources, the commitment is there. The energy team will need to develop a plan, implement it, and keep management informed at all major milestones.

Pitfall 2: Making the entire implementation too technical or too rigid; thus, most people will not understand what is going on or how they can help.

Countermeasure 2: Any training, posters, or information to be given to the employees and contractors of the organization should be reviewed by a small cross-functional test group for review, comments, and recommendations. These individuals can serve as your focus group, ensuring all information provided to the employees and contractors is clear, understandable, and useful.

Pitfall 3: Spending all your time on the technical aspects and little on the EnMS management areas.

Countermeasure 3: Develop a plan that includes all standard elements and phases and implement it in a timely and effective manner.

Pitfall 4: Maintaining two systems—one for everyday use and another for external auditors.

Countermeasure 4: Maintain one system and make sure documents are accurate, legible, and useful.

Pitfall 5: Not keeping everyone informed on what is going on and how they can help; not answering any complaints or cries for information.

Countermeasure 5: Develop an energy awareness training program using Microsoft PowerPoint and determine the best way to provide the training. If management approves, the training may be presented to personnel either by functional area or as a hands-on presentation with all personnel in attendance. A cheap but effective method is to e-mail the training materials to all employees and have them confirm when they finish reading them. This can then be documented for future energy standard audits. At least once a year, preferably every six months, develop a presentation for all employees that tells them of the progress, the results, and the things they can do to help. An energy conservation presentation can be included in this update as well as in the energy awareness presentation.

Pitfall 6: Expecting results almost immediately and not giving O&Ts and projects time to work.

Countermeasure 6: When developing O&Ts, set the review milestones at times when results should materialize. This could be a team review, a project review, or a management review.

Pitfall 7: Not having sufficient resources to support the system.

Countermeasure 7: Insufficient resources can be a real problem when high-cost projects are needed to achieve objectives. Determine what the payback period is for projects. If it is less than three years, there is a good chance that management will fund the project. If the period is more than three years, funding for the project will depend on how committed management is to meeting the corporate energy reduction goal. The energy team should look first at measures that require few resources, such as energy conservation and IT power management.

Pitfall 8: People not convinced and think this is just another management fad and it will go away shortly.

Countermeasure 8: Include in the up-front energy awareness training that the EnMS is a journey, not a destination. Energy usage is an ongoing situation, and attempting to reduce our energy usage and costs is an ongoing process. This standard will not go away, nor will the high energy bills for the many organizations that do not address their energy usage.

Pitfall 9: Not documenting savings, making it difficult to justify later resource expenditures to the board of directors, stakeholders, and others.

Countermeasure 9: This mistake is common. Document all savings. If your organization's top management changes, you will need documentation of the savings in order to justify any later expenditures. Also, documentation is mandatory for auditors to review. Using a centralized system for documentation such as Microsoft SharePoint is a best practice.

Pitfall 10: Not making results visible, due to commingling. For example, new T5 lights were installed at the same time that two new 10-ton air conditioners were installed in a storage room and a new training room. The energy usage actually went up since the lights and the air conditioners were on the same electric meter.

Countermeasure 10: Unless an organization has submeters, some commingling will occur and prevent validation of an energy countermeasure. To verify that reductions have occurred, have an electrician read the volts and amps being drawn to the system before the replacement is made and after (i.e., an electrician takes a reading before T12 lights are replaced with T5 lights and then takes another reading after the replacement is made).

Pitfall 11: Setting goals at the beginning that are overly ambitious and cannot be achieved in the time frame selected with the resources funded.

Countermeasure 11: Make stretch goals that are feasible and can be achieved. Adjusting goals is embarrassing for top management. The energy team should evaluate any countermeasure or project for how many kW or kWh (if an electricity project) will be saved. Ensure that sufficient countermeasures are identified to achieve a goal or target.

Pitfall 12: Not having sufficient reviews to catch problems and correct them before an external audit.

Countermeasure 12: The reviews should include, at a minimum, responsible person reviews, energy team reviews, energy champion reviews, and management reviews. For large projects, peer group reviews and project management reviews should be added.

Pitfall 13: Not following the energy policy in developing new O&Ts after the initial batch was established.

Countermeasure 13: The energy policy should point the way for the type of O&Ts achieved. When establishing O&Ts, several items need to be considered; the energy policy is one of them. Other items include SEUs, recent energy audits, suggestions from the energy champion and top management, team ideas, and metrics.

Pitfall 14: Trying to run the EnMS strictly from headquarters and not keeping the facilities informed and engaged.

Countermeasure 14: If facilities personnel are located close to headquarters, it is good to have some of these individuals on the corporate energy team. If not, consider having a corporate energy team to establish corporate goals, O&Ts, the energy policy, and procedures such as documentation, correction of deficiencies, EnMS self-inspections and audits, management reviews, and others. Then have an energy team for each facility. These teams would identify and establish O&Ts, identify SEUs, and monitor processes and projects specific to their facility. Communication between facility teams and the corporate energy team is a must.

Pitfall 15: Not having sufficient stratification and analysis of an SEU's components.

Countermeasure 15: The stratification of SEUs must go down to individual parts or components. Once stratification is done, decisions on what parts or components need to be replaced can be made.

Pitfall 16: Not having adequate metering to help focus on the real energy eaters.

Countermeasure 16: Unfortunately, most organizations have electric or natural gas meters that measure what enters the facility. They do not have submeters. Thus, the amount of kWh used for lights, for air-conditioning or heating, for office machines, and for plant equipment cannot be determined. Additionally, an organization with a large data center cannot determine how much energy the data center uses compared with the other buildings. For these reasons, a good submeter plan should be identified and funded over time so proper evaluations can be achieved.

Pitfall 17: Making process changes and not retraining process operators.

Countermeasure 17: A before-and-after process flowchart should be developed for any process changes. Personnel who work in the process and/or supervise it must be trained on the changes and what is now expected of each of them. Process objectives, decision points, process metrics and outcome metrics, personnel activities, and inspections are a few of the items that need to be covered in the training.

Pitfall 18: Not making action plans detailed enough; thus, when the action plans were implemented and completed, the target was not achieved.

Countermeasure 18: When developing the activities or actions needed to accomplish a target and showing both who will accomplish it and when, ask the question, "If all the actions, activities, and steps are completed on time, will the target be achieved?" If not, then changes, including adding more or different activities, are needed.

Pitfall 19: Not writing CARs even though nonconformances have been found. The team is afraid of repercussions from management and wants to paint a "rosy" picture.

Countermeasure 19: All deficiencies or nonconformances must be documented so that a record of them is available and tracked to correction. Proper personnel should be notified both when the problem is identified and when it is resolved or eliminated.

Pitfall 20: Not following through with O&Ts or projects for identified opportunities.

Countermeasure 20: Identified opportunities resulting from an affinity diagram, a brainstorming session, an energy audit, or other source should be placed in meeting minutes or attached to the minutes and reviewed each time O&Ts are established. This practice should continue until the item is either selected for O&T or withdrawn since it is no longer valid.

Pitfall 21: Not keeping EnPIs current and not making them visible to all involved. Interest wanes.

Countermeasure 21: Measures showing the status of energy reduction should be placed on bulletin boards, in organization newspapers, or anywhere that employees and contractors can view them. They should be current and easy to read and understand.

Pitfall 22: Not continuing onward after achieving some early success. Top management, the energy champion, or the energy team has not bought into continuous improvement.

Countermeasure 22: Remember, the EnMS is a journey. It should not end when early success has been achieved. Celebrate, but continue to implement the EnMS and continue to improve.

Pitfall 23: Performing self-inspections without checking each element thoroughly, thus painting a false picture.

Countermeasure 23: Self-inspections must be done thoroughly and honestly and must consider all elements. Otherwise, the value of the self-inspection is lost, and the organization will not be ready for an outside audit.

Pitfall 24: Telling management what you think it wants to hear and not the reality.

Countermeasure 24: The energy team should determine what management should be told. Members of top management should be told the truth and the facts so that they can prepare for any problems or occurrences that could arise later.

Pitfall 25: Not giving team members sufficient time to fully implement all the requirements of the standard. Team members are part-time on the energy team and have their normal duties competing for their time.

Countermeasure 25: This is a problem when the team is first formed. But as the team achieves success, management will become more committed and will ensure the team is given sufficient time. If the team is not given sufficient time, either the team leader or the energy champion can talk with the managers of team members and ask that these individuals have more time allotted for energy team matters.

Pitfall 26: Too many layers of the organization to go through to get a procedure approved, changed, or deleted.

Countermeasure 26: The management representative needs to highlight the problem to higher management, showing the flow and the time needed for the process. Highlight the bottlenecks. Recommend a shorter process that includes only the offices and individuals who are directly involved.

PITFALL OBSERVERS

It is the responsibility of the energy champion, the energy team leader, and the team facilitator to observe whether the effort is headed toward one of the pitfalls. If so, corrective action must be initiated immediately by implementing the appropriate countermeasure.

Chapter 10
Implementing ISO 50001 EnMS in Four Months or Less

INTRODUCTION

In 2012, ISO 50001 EnMS was implemented by European University in four months. Any organization that has top management committed to planning, developing, and implementing an EnMS can do so in four months or less but must wait additional time for all the results to materialize. This astounding feat can be done if this book is used to guide the energy champion and the energy team. Otherwise, the normal planning, development, and implementation time is around 12–13 months. After that, maintenance begins that must be continued and sustained.

PROPOSED SCHEDULE

The activities in the recommended time frame are shown in Table 10.1. The start date is May 1, 2014, and completion is scheduled for August 31, 2014 (four months in duration). To review, the different stages of the EnMS are planning and development, implementing, maintaining, and sustaining. The proposed schedule shown in Table 10.1 covers only planning and development, and implementing.

As the schedule shows, by using this book as a guide, ISO 50001 EnMS can be implemented in a short time. If ISO 50001 EnMS is integrated with ISO 14001 EMS and/or ISO 9001 QMS, it can be accomplished in an even shorter time period. What is important is that by implementing the standard, the organization will save energy, reduce GHGs, reduce costs, and have a productive and efficient staff who can and will take on future challenges, regardless of what they are.

Table 10.1 Proposed implementation schedule.

Checklist activity	May 2014	June 2014	July 2014	August 2014	Remarks
Getting started					
1. Gain management commitment	First month				Already have commitment if they set an energy goal, appointed an energy champion, or chartered a team
2. Appoint a management representative/energy champion	First month				
3. Charter a cross-functional energy team	First month				
4. Establish deployment organization	First month				Top management, energy champion, energy team, facilities
5. Have energy champion and energy team review ISO 50001 EnMS and use as a guide	First month				Can be done on own time as homework
6. Develop energy policy and communicate to all employees	First month				Energy team develops draft, energy champion fine-tunes, and top management approves
7. Start team meetings using PAL and PAPA	First month				PAL: purpose agenda limited (time frame) PAPA: purpose, agenda, points (made at meeting), action items (who, what, and when)
Planning and energy review					
8. Develop legal and other requirements	First month				Use Google or ASK
9. Set up a central file system to keep documents	First month				See files recommended
10. Perform an energy review		Second month			Use process in Chapter 4
11. Identify SEUs		Second month			
12. Determine EnPIs, graph baseline and targets	First month				
Objectives and targets					
13. Consider possible technology innovations		Second month			

Table 10.1 Proposed implementation schedule. *(Continued)*

Checklist activity	May 2014	June 2014	July 2014	August 2014	Remarks
14. Consider financial condition of organization		Second month			Talk with comptroller
15. Establish O&Ts		Second month			At least three to five O&Ts
16. Develop action plans for each O&T		Second month			
17. Identify possible reduction projects or efforts			Third month		
18. Estimate contribution and rate of return or payback period			Third month		
Develop and implement					
19. Develop a measurement and monitoring plan		Second month			
20. Develop a communication plan and implement		Second month			Use QVS's as a guide
21. Develop and administer energy awareness and competency training		Second month			Use QVS's as a guide
22. Develop operational controls for each SEU			Third month		Use QVS's as a guide
23. Implement objectives and targets			Third month		
24. Monitor EnPIs and analyze energy variables			Third month		
25. Develop and communicate a contingency plan			Third month		Use QVS's as a guide
26. Develop and communicate an energy saving procurement plan			Third month		
27. Manage and control all documents	First month	Second month	Third month	Fourth month	Every month
Checks					
28. Perform checks (responsible persons, operations manager, team, energy champion)	First month	Second month	Third month	Fourth month	Every month

(continued)

Table 10.1 Proposed implementation schedule. *(Continued)*

Checklist activity	May 2014	June 2014	July 2014	August 2014	Remarks
29. Perform self-inspections and correct deficiencies					Required by May 2015
30. Evaluate legal and other requirements					Required by May 2015
31. Write and resolve CARs and PARs where needed					Whenever deficiency or nonconformity is discovered
32. Get a second-party audit every three years					First is in 2017
Continuous improvement					
33. Maintain the required basics					Continuous
34. Take corrective actions when needed					Whenever needed
35. Comply with the requirements of ISO 50001 EnMS standard					Continuous
36. Sustain the EnMS and use PDCA					Not a destination but a journey

Appendix A
QVS Corporation Management Review

QVS Corp. Management Review
Jan. 25, 2013

A Management Review of QVS Development and Implementation of ISO 50001 EnMS

QVS Corp. Management Review
Jan. 25, 2013

- QVS Corp.'s Power Down Program

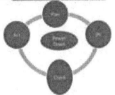

Management Review Agenda
January 25, 2013

- 09:00-09:05 Introductions & Mgt. Review Requirements — Team Leader
- 09:05-09:08 Legal Requirements & Other Facts — Team Leader
- 09:08-09:10 Self Inspection & Corrective Actions — Team Leader
- 09:10-09:15 QVS Energy Policy — Team Leader
- 09:15-09:20 EPIs (Energy Performance Indicators) — Team Leader
- 09:20-09:25 EPIs' Baseline Data & EI Performance — Team Leader
- 09:25-09:30 Completed Objectives & Targets(O & Ts) — Team Leader
- 09:30-09:40 Current O & Ts — Objective Responsible Persons
- 09:40-09:43 Completed Projects & Current Projects — Team Leader
- 09:43-09:45 Changes in Energy Policy & EPIs — Team Leader
- 09:45-09:50 Planned Future Actions — Team Leader
- 09:50-10:00 Questions, Answers, Approvals, & Recommendations — Objective Champion

---

ISO 50001 EnMS Management Review Requirements–Inputs

- Energy management action plan reviews, energy diagnoses/ review results, energy management system audits results; (this includes changes to EPIs).
- Evaluation of legal and other compliance and any changes to legal requirement.
- The energy performance of the organization (how are we doing relative to EPIs).
- The status of corrective and preventive actions.
- The performance of the management system for energy.
- The extent to which energy objectives and targets have been met.
- Recommendations for improvement.

ISO 50001 EnMS Management Review Requirements–Outputs

- The improvement in the energy performance of the organization since the last review.
- Changes to the energy policy.
- Decisions regarding the energy performance of the organization.
- Decisions regarding the energy management system.
- The validity/suitability of EPIs.
- Changes to the objectives, targets or other elements of the management system for energy consistent with the organization's commitment to continual improvement.
- Allocation of resources.

Legal Requirements & Other Facts

- Electricity Provider is Gun Barrel Electric Company. Does not provide a UESCs(Utility Energy Savings Contract) service.
- Can pursue another electric provider since Texas is a deregulated state.
- Must keep power factor above .95 or pay an adjustment fee.
- Must keep all electric conduit and system inside the facilities maintained in workable condition.
- Present cost of electricity is 6.5 cents per KWH.

Appendix A

Corrective Actions

- No Outside or 2nd party audits.
- In Oct 2012, the energy team performed a self energy audit. All discrepancies (primarily in documentation) have been corrected.
- Next self inspection is scheduled for Oct 2013.

QVS Corp. Energy Policy

- QVS Corp. is committed to purchasing and using energy in the most efficient, cost effective, and environmentally responsible manner possible. Therefore, QVS facilities shall:
- Practice energy conservation at all its facilities.
- Lower its peak demand at facilities.
- Improve energy efficiency while maintaining a safe and comfortable work environment.
- Lower our kilowatt hours per square foot to best in class levels. and
- Increase our percentage of renewable energy used. and
- Continuously improve our performance.

Electricity Performance Indicators (EnPIs)

- KWH Usage by Month and Summed for Year
- KWH Usage/Gross Square Footage Summed for Year
- 2010 is our Baseline Year
- QVS Goal is to reduce KWH usage by 10 % by 2015 Year End.

2010 KWH Usage By Facilities & Corp. Baseline

	KWH Usage
Headquarters	2,681,740
Plant A	3,495,709
Plant B	3,423,075
Plant C	3,275,166
Plant D	3,245,210
QVS Total	16,120,900

Electricity Intensity by Facility and Corp. Baseline

Electricity Intensity (KWH/Sq.Ft.)
- Headquarters 48.76
- Plant A 34.1
- Plant B 27.38
- Plant C 33.42
- Plant D 32.45
- QVS 33.55

Present EPI Performance Versus Baseline (2012 vs 2010)

- The Trend has been lower with a 3% reduction from 2010 baseline to end of 2012.

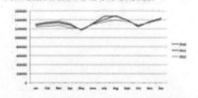

Present EPI Performance Versus Baseline (2012 vs 2010)

- Ten of the 12 months in Year 2012 KWH Usage was lower than that of Year 2010.

Actual Versus Targets/Goal

- QVS Goal is to reduce KWH usage by 10 % by 2015 Year End.
- Results At end of Year 2011   0.16 %
  - At end of Year 2012   3 %
- Targets At End of Year 2013   5%
  - At End of Year 2014   7%
  - At End of Year 2015   10%

Completed Objectives and Targets

- OT-11-01 Energy Awareness Training-In 2011, all employees and contractors have been trained. Will retrain in 2014.
- OT-11-02 Energy Competency Training-Facility Managers were also made the Energy Managers. An Energy Manager Training Course was developed and presented by the team leader and facilitator that included their roles and responsibilities, Basic terms, the EPIs, Things to look for to achieve savings, the Energy Conservation Program and how they can help.
- OT-11-03 Communication Plan-A communication plan has been developed to show how communications on energy should be conducted both in house and external to the organization. All employees have been trained on its contents.
- OT-11-05 Energy Conservation Program-An energy conservation program was developed, put into power point, and sent to all employees and contractors by the objective champion in late November 2011.

Current Objectives and Targets

- OT-11-04 Develop and implement a Purchase Control Plan-Actions revised to include re-negotiating with present provider, and others available due to deregulation, to lower the present cost of 6.5 cents per KWH. In addition, QVS wishes to raise the percentage of renewable energy from present 2% to 15%.
- OT-12-01 Implement an Information Technology (IT) Power Management Program. Have enabled the sleep function on all computer monitors, enabled the "hibernate" function on 25% of the computers, and 10% on all laptops. Ensuring that all new electronics purchases are Energy Star. Recycling end of life electronics with a R2 electronic recycler.

Completed Projects

- **Hqs**-T-12 to T-5 Lights Completed on Nov 30, 2011.
  Plant A-T-12 to T-5 Lights Completed on Dec 10, 2011.
  Plant B-T-12 to T-5 Lights Completed on Dec 30, 2011.
- **Hqs, Plants A-D**-Installed 120 occupancy sensors in areas of infrequent use (rest rooms, break rooms, mechanical rooms, & copier rooms)
- **Plants A,C&D**-Installed 12 inches of insulation in the roof ceilings.
- **Plant D**-Replace Exit signs with low energy use signs.

Current Projects

- **Hqs Building Data Center**-Install panels to separate hot air from servers from the cold air. 30 % Completed. Estimated Completion Date is April 30, 2013
- **Plants A,B,C & D**-Install an advanced electricity meter at the facility. 10% complete. Estimated completion date is Feb 2, 2013.
- **Plant B & D**-Replace doors or fix air leaks. 10 % complete. Estimated completion date is Feb 15, 2013.
- **Plant B**-Install Big As. Fans-20% complete. Estimated completion date is March 30, 2013.
- **Plant D**-Replace security lights with low energy use lights. 10% complete. Estimated Completion is Nov. 30, 2013

Future Projects/ECMs

- Replace Cooling Tower in Hqs Building. Pay Back Period is 8.2 Years.
- Replace Boiler in Plant B Pay Back Period is 9.5 Years.
- Replace Hot Water Heaters in Hqs Building, Plant A,B & D with Solar Panels. Pay Back Period is 3.7 years.

Changes in QVS Corp. Energy Policy?

- **QVS Corp. is committed to purchasing and using energy in the most efficient, cost effective, and environmentally responsible manner possible. Therefore, QVS facilities shall:**
- Practice energy conservation at all its facilities.
- Lower its peak demand at facilities.
- Improve energy efficiency while maintaining a safe and comfortable work environment.
- Lower our kilowatt hours per square foot to best in class levels. and
- Increase our percentage of renewable energy used and
- Continuously improve our performance.
  Green is on track. Red means we need to address.

Energy Policy Recommendations

- Recommendations:
  #1. Keep Energy Policy as is.
  #2. Recommend a new objective and target be approved to address peak load.
  #3. Get Strategic Council to approve 15 % by end of 2015 as our renewal energy goal or target.

EPI Changes?

- Keep our EPIs: KWH Usage by Month and Summed for Year
- KWH Usage/Gross Square Footage Summed for Year
- Recommendation # 4: Keep EPIS above. They are valid and suitable.
- Recommendation # 5: For the objective and target to reduce the peak load, recommend we add Electric Load Factor to be calculate quarterly to ensure our peak loads have not changed negatively for some reason. (see definition of peak load and the ELF on next three slides.)

PEAK DEMAND

Peak Demand is the highest demand over the demand period in the billing period (normally a calendar month). In other words, it is the highest amount of power your facility requires at a given time. Often this leads to Demand Charges. You can think of demand charges as overhead expenses that your utility incurs to provide the electricity infrastructure that is capable of meeting your largest electric load.

ELECTRIC LOAD FACTOR

- Electric Load Factor (ELF) is an indicator that shows if peak demand is high for your facility. It is an indicator of how steady an electrical load is over time. The optimum load factor is 1 or 100%. The closer to zero, the more you are paying for electricity.

- ELF (%)=Total KWHs/#Days in Electricity Bill Cycle X 24 Hours/Day/ Peak KW Demand.

In other words, it is the average demand/peak demand for a given period of time. From your electricity bill get the KWHs used and the peak KW. Next look for days included in the bill. Multiply these days by 24(hours per day). Divide this number into the KWHs. Then divide what you get by the peak KW. Multiply this number by 100 to get into percentage.

ELECTRIC LOAD FACTOR

- For example: If a facility used 125,000 KWH in July where the billing period covered 30 days, The peak KW demand was 218, the ELF is calculated by (125,000/30x 24/218) x 100=79.64%

- Action should be initiated to increase load factor when you are 60% or lower. If low, shift electricity intensity processes to other times. By increasing load factor, you will reduce the impact of monthly demand (KW) charged on your electric bill.

- This calculation only needs to be done quarterly to ensure ELF has not significantly.

Planned Future Actions

- Continue Meeting Monthly even if only 30 minutes duration.
- Manage objectives and targets and projects/ ECMs to completion.
- Standardize the Objective Champion's quarterly update to Strategic Council.
- Continue to identify ECMs.
- Perform 2nd Self Inspection in October 2013.

NEED YOUR APPROVAL & RECOMMENDATIONS

- 1. Approve Current Objectives and Targets
- 2. Recommend recommended new Objective and Target
- 3. Approve other four recommendations.
- 4. Approve future planned actions.

QUESTIONS AND ANSWERS

1. Does the EMS Team's focus meet with your approval?
2. Do the planned future actions meet with your approval?
3. Do you approve our Current O&Ts?
4. Is there anything the EnMS team is not doing, that you would like us to be doing?

Appendix B
List of Acronyms

BAS: building automation system

BAT: best available technique

Btu: British thermal unit

CAR: corrective action report

CPM: critical path method

CRAC: computer room air conditioner

CSF: critical success factor

ECM: energy conservation measure

ELF: electric load factor

EMS: Environmental Management System

EnMS: Energy Management System

EnPI: energy performance indicator

EPA: Environmental Protection Agency

EPEAT: energy performance environmental attributes tool

GHG: greenhouse gas

HDD: heating degree day

HVAC: heating, ventilation, and air-conditioning

ISO: International Organization for Standardization

IT: information technology

LEA: lean energy analysis

LEED: leadership in energy and environmental design

kW: kilowatt

kWh: kilowatt-hour

MSDS: materials safety data sheet

MTBF: mean time between failures

MTTF: mean time to failure

MTTR: mean time to repair

NEPA: National Environmental Policy Act

NGPA: National Gas Policy Act

OAR: objective agenda restricted

OEP: occupant emergency plan

O&T: objective and target

PAL: purpose agenda limited (for meetings, develop a purpose, have an agenda, and limit the time per agenda topic)

PAPA: purpose, agenda, points, action items

PAR: preventive action report

PDCA: plan, do, check, act (the Deming cycle)

PUE: power utilization effectiveness

PURPA: Public Utility Regulatory Policies Act

QMS: quality management system

ROI: return on investment

SEU: significant energy user

SMART: specific, measurable, actionable, relevant, and time-framed

SWOT: strengths, weaknesses, opportunities, and threats

UPS: uninterruptible power supply

Endnotes

CHAPTER 1

1. US Energy Information Administration, http://eia.doe.gov.

2. "Green Building," *Wikipedia*, last modified December 17, 2013, http://en.wikipedia.org/wiki/Green_Building/.

3. International Organization for Standardization, *Win the Energy Challenge with ISO 50001* (Geneva, Switzerland: International Organization for Standardization, 2011), http://www.iso.org/iso/iso_50001_energy.pdf.

4. International Organization for Standardization, ISO 50001 Energy Management, http://www.iso.org/iso/home/standards/management-standards/iso50001/.

5. "ISO 50001," *Wikipedia*, last modified October 29, 2013, http://en.wikipedia.org/wiki/ISO_50001/.

CHAPTER 2

1. *Dallas Morning News*, page 2A, Tuesday, Aug. 15, 2006.

CHAPTER 3

1. "Energy Policy," *Wikipedia*, last modified April 24, 2013, http://en.wikipedia.org/wiki/Energy_Policy/.

CHAPTER 4

1. B. H. Minor and D. B. Bivens, *Reduced Buildings Greenhouse Gas Emissions with Integrated Approach to Design, Systems and Operation* (International Refrigeration and Air Conditioning Conference, School of Mechanical Engineering, Purdue University, 2002), http://docs.lib.purdue.edu/cgi/viewcontent.cgi?article=1529&context=iracc.

2. "Efficient Building Equipment and Systems," Johnson Controls, http://www.makeyourbuildingswork.com/building-management-systems/.

3. US Energy Information Administration, Frequently Asked Questions: "How much electricity is used for lighting in the United States?" http://www.eia.gov/tools/faqs/faq.cfm?id=99&t=3.

4. American Society of Heating, Refrigerating and Air-Conditioning Engineers, ANSI/ASHRAE Standard 100-2006, "Energy Efficiency in Existing Buildings."

5. "Electrical Ballast," *Wikipedia*, last modified December 19, 2013, http://en.wikipedia.org/wiki/Electrical_ballast/.

6. "Light Fixture," *Wikipedia*, last modified January 14, 2014, http://en.wikipedia.org/wiki/Light_fixture/.

7. Leah Garris, "Lighting: The Clean Factor," *Buildings*, January 1, 2009. http://www.buildings.com/article-details/articleid/6833/title/lighting-the-clean-factor.aspx.

8. Natural Resources Defense Council, "Green Advisor: Reducing Paper Use," http://www.nrdc.org/enterprise/greeningadvisor/pa-reducing.asp; Reduce.org, "Reducing Waste in the Workplace," http://reduce.org/workplace/index.html.

9. Efficiency Vermont, "Guide to Savings: Compressed Air Systems," 2011, http://www.efficiencyvermont.com/docs/for_my_business/publications_resources/Compressed_Air_Guide_To_Savings.pdf.

10. Technochem Consultants, Sample Preventive Maintenance Manual: Packaged Fire Tube Boiler, http://technochemconsultants.com/SAMPLE%20PREVENTIVE%20MAINTENANCE%20MANUAL%20FOR%20A%20PACKAGED%20FIRE.pdf.

11. Madison Gas and Electric, "Managing Packaged Rooftop Units," http://www.mge.com/saving-energy/business/bea/_escrc_0013000000DP22YAAT-2_BEA1_OMA_OMA_HVAC_OMA-01.html.

12. Saturn Resource Management, "Cleaning Condenser Coils," http://blog.srmi.biz/energy-saving-tips/residential-air-conditioning-aircon-ac/cleaning-condenser-coils/.

13. John Seryak and Kelly Kissock, *Lean Energy Analysis: Guiding Industrial Energy Reduction Efforts to the Theoretical Minimum Energy Use* (ACEEE Summer Study on Energy Efficiency in Industry, 2005), http://aceee.org/files/proceedings/2005/data/papers/SS05_Panel01_Paper14.pdf.

14. Kelly Kissock and John Seryak, *Lean Energy Analysis: Identifying, Discovering, and Tracking Energy Savings Potential* (Society of Manufacturing Engineers: Advanced Energy and Fuel Cell Technologies Conference, 2004), http://academic.udayton.edu/kissock/http/Publications/LeanEnergyAnalysisSME2004.pdf.

15. EnergyRight Solutions, "Managing Energy Costs in Data Centers," 2009, http://www.energyright.com/business/pdf/Data_Centers_ESCD.pdf.

16. North Carolina Department of Environment and Natural Resources, "Vending Machines: Energy Saving Fact Sheet," http://portal.ncdenr.org/c/document_library/get_file?uuid=cdc48d4d-8e46-451d-9e7b-f2f4bf514829&groupId=38322; Tufts Climate Initiative, "Vending Misers: Facts and Issues," http://sustainability.tufts.edu/wp-content/uploads/VendingMiserHandout-updated020310.pdf; Appropedia, "Motion Sensor Controlled Vending Machines," http://www.appropedia.org/Motion_sensor_controlled_vending_machines.

17. "Heating Degree Day," *Wikipedia*, last modified December 17, 2013, http://en.wikipedia.org/wiki/Heating_degree_day/.

Glossary

action plan—The steps that must be taken to accomplish an objective and target or project showing who is going to do what and when. A Gantt chart is the most common form of an action plan.

analysis—Using Six Sigma tools such as a fishbone diagram or a histogram to find the root cause and then verify it with data.

base load analysis—A method to determine the minimum amount of electricity to be furnished in a time period.

benchmarking—A method that enables one organization to compare its process with the process of an organization whose performance is much better and then use the information to improve its process performance.

best available technique—"Best" means most effective, "available" means it can be used in your situation at a reasonable cost, and "technique" refers to what technology is used and the way in which installation or implementation is designed, installed, maintained, operated, and later decommissioned.

best practice—An accepted way done by a world-class organization that is proven to be the best way of accomplishing something.

brainstorming—An idea-gathering technique that uses group interaction to generate as many items as possible in a designated time. Quantity of ideas is the focus.

charter—A written commitment by management detailing an improvement team's authority and resource needs, including time and funds, to complete an assignment.

checklist—A tool that outlines key events to make something happen.

check sheet—A quality tool designed to collect data.

conformance—A product or service that has met the requirements of the relevant specifications, contract, or regulation; also the state of meeting the requirements.

contribution—The estimated amount that a project or action will reduce the electricity or natural gas usage or any other energy use.

cost-benefit analysis—Analysis to show both the costs and the resulting benefits of a plan.

countermeasure—A potential solution that should eliminate or minimize a root cause to a problem.

critical success factors—What you are paid to accomplish; the process or program will not succeed unless these factors are done properly.

cross-functional—A term used to describe people from several different organizational units or functions brought together on a team for a specific purpose.

cultural change—A major shift in attitudes, norms, sentiments, beliefs, values, operating principles, and behavior of an organization.

customer—Anyone for whom an organization provides goods or services; can be internal or external to the organization. The next user of a product.

CUSUM (cumulative sum)—A technique described in the standard for analyzing energy data that compares the actual consumption with the baseline over the baseline period to determine actual reduction amounts.

defect—A nonconforming attribute.

discrete data—Data you can count, such as number of errors or defects.

DMAIC—A problem-solving process used in Six Sigma: define the problem, measure the progress and results, act to implement actions for improvement, improve, and control the process so the gain is maintained.

effect—An observable action, result, or evidence of a problem. It is what happens.

80-20 rule—80% of the potential savings come from only 20% of the problems.

energy audit—An audit of the facility's infrastructure, building envelope, HVAC, equipment, motors, exit signs, lighting, and so forth, to identify areas where energy use can be reduced. The result is a list of projects with the return on investment for each.

energy champion—The management representative required by ISO 50001 EnMS.

energy conservation measure—A government form that shows projects that, if funded, would reduce energy and/or water use and provides a description of the project and the payback period or the return on investment.

energy intensity—The energy consumed in kBtu divided by the total gross square footage of the facility. This measure enables comparisons with other similar facilities.

energy mapping—A procedure for stratifying main energy users into their components and subcomponents; for example, HVAC components are heating, ventilation, and air-conditioning.

energy model—Recording all energy use on a data sheet and analyzing it to determine where reductions may be possible.

energy product environmental assessment tool (EPEAT)—A certification by a nongovernmental organization that a product is energy friendly and meets the organization's specifications.

energy variables—Variables that are correlated with energy use or can be used to normalize energy use in comparisons with other organizations.

facilitator—An individual who helps a team through a process or method and makes it easier for the team to achieve its purpose.

fat rabbit—The area in which the greatest improvement can be achieved if action is taken.

fishbone diagram—Also known as a cause and effect diagram, this tool was invented by Ishikawa to identify root causes and their effects.

5 Ss—A tool to achieve orderliness and cleanliness that helps improve employee efficiency and productivity. 5 Ss originated at Toyota in Japan and has been implemented all over the world. It is based on the saying "Every thing has a place and everything is in its place." The five Ss explain the process: *Sort* all items in the workplace, identifying those no longer needed (waste); *structure*: develop a structure where everything has a place (e.g., parking a forklift in a designated area at the end of every shift); *shine*: keep the area and items clean; *standardize*: develop standard operating procedures, floor layouts, and so forth, to ensure everything is put in its designated space; *sustain*: build 5S sustainment into the organizational structure and culture.

5 Whys—A technique for discovering root causes of a problem by repeatedly asking and answering the question "Why?"

flowchart—A chart that shows inputs, the process (sequential work activities), outputs, and outcomes.

force field analysis—A graphical representation of barriers and aids to help sell an implementation plan to solve a problem.

goal—A broad statement describing a future condition or achievement without being specific about how much and when. The establishment of a goal implies that a sustained effort and energy will be applied over a time period.

graph—A visual display of quantitative data over time, such as a line chart, column chart, run chart, or bar chart.

input—Materials, equipment, training, people, dollars, energy, facilities, systems, and so forth, that are needed to start a process.

kaizen—Japanese word for "continuous improvement."

kaizen event—Usually a one-day event in which a group works to improve a process.

key results areas—Broad quality requirements where improvement would give the stakeholders the most benefit (i.e., where the biggest bang for the buck can be achieved).

lean analysis—A technique of examining everything in a facility to determine whether it adds any real value. Anything that does not add value should be discontinued.

life cycle costing—The total cost incurred with the purchase and operation of a product over its lifetime.

mean—A measure of central tendency of data (sum of total divided by number of observations); an average.

mean time between failures—Average time between failures for a piece of equipment.

mean time to repair—Average time to repair equipment.

median—The middle point of data. A measure of central data.

metric—Indicator or measure over time of a process whose objective is for improvement. Establishes a standard for management action.

mode—The most frequently recurring number in a data set.

monitoring—Reviewing key indicators or equipment performance at specified times.

multivoting—A technique that allows a group to prioritize a large list of items down to three to five of the most important.

normalize—A method of making energy data or any data comparable with other similar facilities.

objective—A specific statement of a desired short-term condition or achievement; it should include measurable end results to be accomplished within specific time limits. The how and when for achieving a specific goal.

outcomes—Results from outputs; for example, the customer is satisfied, savings have resulted, costs have been avoided, and so forth.

outputs—Products, materials, services, and information provided to customers (internal or external) resulting from a process.

overall equipment effectiveness—A method of breaking a plant, equipment, or a process into availability, performance, and quality to determine their efficiency and effectiveness.

Pareto chart/ABC analysis—A quality tool that identifies the most important or significant problem or opportunity.

PDCA (plan-do-check-act)—The Deming or Shewhart cycle for continuous improvement. The four stages are plan (plan what you are going to do), do (implement the plan), check (check the results), and act (take whatever action is necessary to get back on track to meet the target).

plan—A goal or objective, metric(s), target(s), and a method and activities in which to accomplish it.

policy—A goal, objective, metric(s), and target(s).

power factor—An adjustment fee caused by harmonics in the system inside a facility that forces the utility to have more power than what is used available. The factor goes from 0 to 1, and in some states, anything under 0.94 incurs an adjustment fee on the facility's electric bill.

power management—A way to save energy by placing computers, monitors, and laptops in a sleep mode when they have not been used for a specified length of time.

preventive maintenance—Routine maintenance that is performed to prevent equipment deterioration or failure.

problem-solving process—A structured process that, when followed, will lead to a solution to a problem. DMAIC is an example.

process—A sequence of activities that produce a product, a service, or information.

process capability ratio—Relates customer requirements to actual process performance and shows whether the process is capable of meeting the customers' needs.

process control—A process is under statistical control.

process control system—A system that identifies who does what (flowchart) to achieve an objective of a process. Also identifies process metrics and outcome metrics.

process map or flowchart—A diagram that shows inputs and the sequence of activities to turn the inputs into outputs such as a product, service, or information.

process owner—Individual responsible for managing and controlling the process.

product—A tangible output.

quality—Conformance to valid requirements.

quality assurance—A process of obtaining quality by measurement and analysis of work methods, building quality into the design phase, not inspecting out.

quality improvement—A systematic method for improving processes to better meet a client's needs and expectations.

quality system—A chart that depicts what activities or processes, along with the necessary information and data, a company must undertake to ensure quality and effectiveness at each step.

quality tool—An instrument or technique that supports the activities of process quality management and improvement.

range—The highest point of data minus the lowest.

regression analysis—A technique that allows for determining one variable's relationship with another variable or variables.

reliability centered maintenance—A method to identify the critical parts of a plant or facility and ensure that they are always operable.

root cause—The reason a requirement in a process is not being achieved.

root cause analysis—Analysis using a technique such as a fishbone diagram to identify the root cause(s) that caused the problem to occur.

self-inspection—An audit by a team using a checklist of the key elements to determine whether they comply with and conform to an ISO standard.

Six Sigma—A statistics-based improvement methodology that strives to limit defects to 3.4 failures per million possibilities or six standard deviations from the mean.

stakeholders—Individuals who have a stake in their organization's mission, changes, and results (e.g., customers, employees, suppliers, management).

standard—Factor usually established through statistics or measurement that serves as a basis for comparison.

strategic direction—A vision of where an organization wants to go in the future or what it desires to be.

strategic objective—A corporate objective that supports the vision and strategic direction established by the organization.

strategy—A determination of how you impact an objective to get where you want to be.

system boundary—A statement of what elements are included under the analysis being accomplished.

target—A desired state or standard tracked through an indicator or the method or strategy that an objective is going to achieve.

task—A specific, definable activity to perform an assigned piece of work, often finished within a certain time frame.

team—A collection of people brought together to work on an objective or problem.

team leader—A person who leads a team through a process to achieve an objective or solve a problem.

team member—An individual who serves on a team.

teamwork—Harmoniously working as a group to achieve an objective or solve a problem.

training needs analysis—An analysis of each person's training requirements, often required by a standard.

value added—What happens in a process to change the inputs into outputs.

vision—Through analysis and customer voice, senior management determines where the organization should be in the long term (5–10 years).

"walk the talk"—Leading by setting a good example.

Bibliography

BOOKS/ARTICLES/STANDARDS

Boynton, A. C., and R. W. Zmud. "An Assessment of Critical Success Factors." *Sloan Management Review* 25, no. 4 (1984): 17–27.

Camp, Robert C. *Benchmarking: Finding and Implementing Best Practices That Lead to Superior Performance.* Milwaukee, WI: ASQC Quality Press, 1989.

Chang, Kae H. *Management: Critical Success Factors.* Weston, MA: Allyn and Bacon, 1987.

Denning, S. "How Do You Change an Organizational Culture?" *Forbes*, July 23, 2011. http://www.forbes.com/sites/stevedenning/2011/07/23/how-do-you-change-an-organizational-culture/.

Howell, Marvin T. *Actionable Performance Measurement: A Key to Success.* Milwaukee, WI: ASQ Quality Press, 2005.

———. *Critical Success Factors Simplified: Implementing the Powerful Drivers of Dramatic Business Improvement.* Boca Raton, FL: CRC Press, 2010.

Huotari, M.-L., and T. D. Wilson. "Determining Organizational Information Needs: The Critical Success Factors Approach." *Information Research* 6, no. 3 (2001):

ISO 14001 Environmental Management System Standard, Requirements with Guidance for Use, 2004, ISO Central Secretariat.

ISO 50001 Energy Management System Standard, Requirements with Guidance for Use, 2011, ISO Central Secretariat.

ISO Energy Management Systems, Guidance for the Implementation, Maintenance, and Improvement of an EnMS, October 28, 2011, ISO/WD.2, Secretariat: ANSI.

Mears, Peter. *Quality Improvement Tools and Techniques.* New York: McGraw-Hill, 1995.

Mizuno, Shigeru, ed. *Management for Quality Improvement: The Seven New QC Tools.* Cambridge, MA: Productivity Press, 1988.

Randolph, John, and Gilbert M. Masters. *Energy for Sustainability: Technology, Planning, Policy.* Washington DC: Island Press, 2008.

Rockart, John F., and Christine V. Bullen. "A Primer on Critical Success Factors." Center for Information Systems Research, Sloan School of Management, Massachusetts Institute of Technology, 1981.

Shaligram, Pokharel. *Energy Analysis for Planning and Policy.* Boca Raton, FL: CRC Press, 2014.

Stremke, Sven, and Andy Van Den Dobbelsteen, eds. *Sustainable Energy Landscapes: Designing, Planning, and Development.* Boca Raton, FL: Taylor & Francis, 2013.

US Government Accountability Office. "Federal Energy Management: Addressing Challenges through Better Plans and Clarifying the Greenhouse Gas Emission Measure Will Help Meet Long-term Goals for Buildings." September 2008. http://www.gao.gov/products/GAO-08-977.

WEBSITES

Alliance to Save Energy, http://www.ase.org.

American National Standards Institute, http://www.ansi.org.

The Electricity Forum, http://www.electricityforum.com.

International Energy Agency, http://www.iea.org.

International Organization for Standardization, http://www.iso.org.

Sustainable Energy Authority of Ireland, Energy Management Action Programme (MAP), http://www.seai.ie/energymap/.

US Department of Energy, Energy Management System (EnMS) Implementation Self-Paced Module ("Toolkit"), http://www1.eere.energy.gov/energymanagement/pdfs/self-paced_module_outline.pdf.

US Department of Energy, Federal Energy Management Program: Training, http://apps1.eere.energy.gov/femp/training/index.cfm.

US Department of Energy, Office of Energy Efficiency and Renewable Energy, http://energy.gov/eere/office-energy-efficiency-renewable-energy/.

Index

Note: Page numbers followed by *f* refer to figures; those followed by *t* refer to tables.

A

acronyms, list of, 161–162
action plans and projects, energy planning, 60–63, 61–63*t*, 61*f*
air compressors, energy efficiency and, 39
air filters, energy efficiency and, 39
awareness, 66–67

B

bearings, energy efficiency and, 39
belts, energy efficiency and, 39
best available technique (BAT), 118
boilers, energy efficiency and, 39
British thermal unit (Btu), 35, 48–49
building automation system (BAS), 37, 61*t*, 62*t*, 63*t*
building envelope, energy efficiency and, 39
building seals, energy efficiency and, 39

C

checking phase, 85–114. *See also* critical success factor (CSF)
    control of records, 114
    data presentation, 86, 90*f*
    described, 85
    EnMS standard, 85
    internal audit, EnMS, 99–100, 100–109*f*
    legal requirements, compliance with, 96, 96*t*, 97–99*f*
    monitoring, measurement, and analysis, 85–96, 87*f*, 88–89*t*, 90*f*, 92–93*f*, 94–95*t*, 95–96*f*
    nonconformities, corrective, preventive actions, 109, 110–113*f*
    operational explanation, 86
    QVS implementation example, 86–90, 87*f*, 88–89*t*, 90*f*
communications, 70–71
    data presentation, 70–71
    EnMS standard, 70
    operational explanation, 70
    QVS implementation example, 71, 71–73*f*
competence, training, and awareness, EnMS, 66–67
    data presentation, 67, 68*t*
    EnMS standard, 66
    operational explanation, 66–67
    QVS implementation example, 67, 69*t*
computer room air conditioner (CRAC), 35*t*, 41
condenser coils, energy efficiency and, 40
control of records, 75–76, 75*f*
    checking phase and, 114
    data presentation, 76
    EnMS standard, 75
    operational explanation, 76
    QVS implementation example, 75*f*, 76
corrective action reports (CARs), 4, 110–111*f*
countermeasures, pitfalls and, 147–152
critical path method (CPM), 51, 81
critical success factor (CSF), 90–96
    assessment of, 91, 94–95*t*
    described, 90
    measurement of, 91, 92–93*f*
    QVS scores, 95–96*f*

D

data centers
    energy efficiency and, 41
    SEU consumption and, 41
day lighting, energy efficiency and, 41
Deming cycle (PDCA), 4, 4*f*, 7
design, 79–81
    data presentation, 80–81, 80*f*, 82*t*
    EnMS standard, 79
    operational explanation, 79–80, 80*f*, 83*f*
    QVS implementation example, 81, 83*f*

documentation, 73–74
   data presentation, 74
   EnMS standard, 73
   operational explanation, 73–74, 75f
   QVS implementation example, 74, 75f
duplex printing, energy efficiency and, 41

E

economizers, energy efficiency and, 41
electric load factor (ELF), 123f
energy, corporate vision for, 21–24
energy baseline, 46–47
   data presentation, 47
   EnMS standard, 46
   operational explanation, 47
   QVS implementation example, 47, 47t
energy conservation measure (ECM), 79–81, 80f
energy cost, 1
energy efficiencies, identifying, 37–40
   air compressors, 39
   air filters, 39
   boilers, 39
   building envelope/seals, 39
   condenser coils, 40
   fans, bearings, and belts, 39
   HVAC system, 37
   leaks, 40
   lighting, 37–38
   motors, 39
   office machines, 38–39
Energy Management System (EnMS). *See also* requirements, EnMS
   demonstration example of implementing, 2–3
   elements of, 3–4, 4f
   energy cost and, 1
   implementation and operation of, 65–84
   integration of, 2
   internal audit of, 99–100, 100–109f
   ISO 50001 definition of, 1
   management representative and, 10–13
   management responsibilities for, 8–10
   phases of, 3–4, 3f
   requirements for, 3, 7–8
   stages of, 4–5, 5f
   standard, 1–2
   team meetings and, effectiveness of, 13–16, 14f, 15–16f, 17–20t
energy objectives and targets, 51–60
   data presentation, 53–56, 54t, 55f
   EnMS standard, 51
   operational explanation, 52–53
   QVS implementation example, 56–60, 57t, 59f
   types of, 52–53
energy performance environmental attributes tool (EPEAT), 38, 45t
Energy Performance Intensity (EnPI), 47–51
   data presentation, 48, 48f
   EnMS standard, 47
   operational explanation, 48
   QVS implementation example, 48–50, 49f, 50–51f
energy planning, 25–63
   action plans and projects, 60, 61–64t, 61f
   baseline, 46–51, 47t, 48f, 49f, 50–51f
   data presentation, 26
   efficiencies, identifying, 37–40
   energy baseline and, 46–47
   energy review, 28–36, 30f, 31f, 31t, 32–34f, 35–36t
   energy variables and, identifying, 43–46, 44–46t
   EnMS standard, 26
   EnPIs and, 47–51
   explained, 26
   general requirements, 26
   lean energy analysis (LEA), 40–42
   legal and other requirements, 26, 27f
   objectives and targets, 51–59, 54–55t, 57t, 59f
   operational explanation, 26
   overview of, 25
   QVS implementation example, 26, 27f
   variables, identifying, 43, 44–46t, 46
energy policy, 21–24
   commitment statement, 22–23, 23f
   data presentation, 22–23, 23f
   defined, 21
   EnMS standard, 21
   operational explanation, 21–22
   QVS implementation example, 23, 23f, 24t
   use of, 22
energy review, 28–36
   data presentation, 29–30, 30f
   EnMS standard, 28
   operational explanation, 28–29
   QVS implementation example, 30–36, 31t, 32–34f, 35t, 36t
Energy Star, 38
energy variables, identifying, 43–46, 44–46t
EnMS (Energy Management System), energy efficiency and, 41

EnMS standard, ISO 50001, 1–2
    general requirements of, 7
    purpose of, 7
EnPI (Energy Performance Indicator), 47–51
    data presentation, 48, 48*f*
    EnMS standard, 47
    operational explanation, 48
    QVS implementation example, 48–50, 49*f*, 50–51*f*
Environmental Management System (EMS), 1, 132*t*
Environmental Protection Agency (EPA), 27*f*, 98*f*

F

fans, energy efficiency and, 39

G

Gantt chart, 81, 82*t*
greenhouse gas (GHG), 37, 129*f*

H

heating degree days (HDDs), 43, 46
HVAC system, energy efficiency and, 37

I

implementation and operations, 65–84
    communications, 70–71, 71–73*f*
    competence, training, and awareness, 66–67, 68–69*t*
    control of records, 75–76, 75*f*
    data presentation, 66
    design, 79–81, 80*f*, 82*t*, 83*f*
    documentation, 73–74, 75*f*
    EnMS standard, 65
    operational controls, 76–77, 77–78*t*, 79*f*
    operational explanation, 65
    overview of, 65–66
    procurement of services, products, and equipment, 84, 84*f*
    QVS implementation example, 66
implementation of EnMS stage, 4, 5*f*, 16
implementation schedule, 154–156*t*
improve type of objective, 52
inputs to management review, 116–118
    data presentation, 117
    EnMS standard, 116
    operational explanation, 116–117
    QVS implementation example, 117–118, 117*f*

internal audit, EnMS, 99–100, 100–109*f*
    data presentation, 100, 100–104*f*
    EnMS standard, 99
    operational explanation, 99–100
    QVS implementation example, 100, 105–109*f*
ISO 9001 Quality Management System, 1, 2, 125, 153
ISO 14001 Environmental Management System, xiii, 1, 2, 125, 129*f*, 130*f*, 134*f*, 153
ISO 50001 EnMS, implementing, in four months or less, 153–156
    introduction to, 153
    schedule for, 153, 154–156*t*
ISO 50001 Environmental Management System (EnMS), 1–5
    energy cost, 1
    EnMS standard, 1–2
    implementing, in four months or less, 153–156
    integration of ISO standards, 2
    phases and elements, 3–4, 3–4*f*
    QVS Corporation (hypothetical company) demonstration use, 2–3
    stages, 4–5, 5*f*
ISO standards, integration of, at QVS, 2, 125–142
    communication plan, 134–136*f*
    corrective action request (CAR), 138–139*f*
    documentation files, 132–133*t*
    elements, evaluation of, 127–128*t*
    emergency preparedness plan, 143*t*
    energy contingency plan, 142*f*
    environmental/energy policy and, 129*f*, 140*t*
    management review agenda, 141–142*f*
    objectives and action plan, 137*f*
    structure used for, 2–3
    summary of, 144–145*t*
    team roles and responsibilities, 130–131*f*
IT power management, energy efficiency and, 40

K

kilowatt (kW), 29, 61*f*, 89
kilowatt-hour (kWh), 27*f*, 31*t*, 32–34*f*, 43, 47, 47*t*, 48–49, 49*f*, 50*f*, 59*f*, 60, 61–63*t*, 121–123*f*, 137*f*

L

leadership in energy and environmental design (LEED), 79–80, 81

leaks, energy efficiency and, 40
lean energy analysis (LEA), 40–42
LEED (leadership in energy and environmental design), 79–80, 81
legal requirements, compliance with, 96–99
    data presentation, 97
    EnMS standard, 96
    operational explanation, 97
    QVS implementation example, 97, 97t, 98–99f
lighting, energy efficiency and, 37–38, 40

M

maintaining EnMS stage, 4, 5f
maintain type of objective, 52
management representative, 10–13
    energy team, 10–11
    EnMS standard, 10
    operational explanation, 10–11
    QVS implementation example, 11–13
management responsibilities, 8–10
    EnMS standard, 8
    operational explanation, 8–9
    QVS implementation example, 9–10
management review, 115–123
    data presentation, 116
    EnMS standard, 115
    general requirements, 115–116
    inputs to, 116–118, 117f
    operational explanation, 115
    outputs from, 118–119, 120–123f
    QVS Corporation example of, 157–160
    QVS implementation example, 116
mean time between failures (MTBF), 46t
mean time to failure (MTTF), 46t
monitoring, measurement, and analysis checking phase, 85–96
    CSF assessment, 90–96, 92–93f, 94–95t, 95–96f
    data presentation, 86, 90f
    EnMS standard, 85
    operational explanation, 86
    QVS implementation example, 86–90, 87f, 88–89t, 90f
motors, energy efficiency and, 39
Murphy's Law, 147

N

National Environmental Policy Act (NEPA), 27f, 98f

Natural Gas Policy Act (NGPA), 27f, 98f
nonconformities or deficiencies, 109–113
    data presentation, 109, 110–111f
    EnMS standard, 109
    operational explanation, 109
    QVS implementation example, 109, 112–113f

O

OAR. *See* objective agenda restricted (OAR)
objective agenda restricted (OAR), 13–15, 15–16f
objectives, types of, 52–53
objectives and targets. *See* energy objectives and targets
occupancy sensors, energy efficiency and, 41
occupant emergency plan (OEP), 143t
office machines, energy efficiency and, 38–39
operational controls, 76–77
    data presentation, 77
    EnMS standard, 76
    operational explanation, 76, 77–78t
    QVS implementation example, 77, 77–78t, 79f
O&Ts (objectives and targets). *See* energy objectives and targets
outputs from management review, 118–119
    data presentation, 119
    EnMS standard, 118
    operational explanation, 118–119
    QVS implementation example, 119, 120–123f
outside-air intake controls, energy efficiency and, 42

P

PAL (purpose agenda limited), 13–15, 15–16f
PAPA (purpose, agenda, points, action items), 15, 154t
PDCA (plan, do, check, act), 4, 4f, 7
phases and elements, EnMS, 3–4, 3–4f
pitfall observers, 152
pitfalls and countermeasures, 147–152
plan-do-check-act cycle (PDCA), 4, 4f, 7
planning and development of EnMS stage, 4, 5f
plug loads, energy efficiency and, 41
power utilization effectiveness (PUE), 48, 62t, 87, 88t
preventive action reports (PARs), 4

process heating, energy efficiency and, 42
procurement plan, energy services, products, and equipment, 84, 84*f*
    EnMS standard, 84
    operational explanation, 84, 84*f*
    QVS implementation example, 84, 84*f*
Public Utility Regulatory Policies Act (PURPA), 27*f*, 98*f*
purpose agenda limited (PAL), 13–15, 15–16*f*
purpose, agenda, points, action items (PAPA), 15, 154*t*

Q

Quality Management System (QMS), 1
QVS Corporation (hypothetical company). *See also* ISO standards, integration of, at QVS
    communication plan, 71–73*f*, 134–136*f*
    contingency plan, 79*f*
    control of records, 75*f*
    corrective action request (CAR), 112–113*f*, 138–139*f*
    CSF scores of, 95–96*f*
    demonstration use, 2–3
    documentation plan, 75*f*
    electricity baseline, 47*t*
    electricity profile, 32–34*f*
    emergency preparedness plan, 143*t*
    energy conservation measure (ECM) and, 81, 83*f*
    energy contingency plan, 142*f*
    Energy Performance Intensity (EnPI), 49*f*, 50–51*f*, 88–89*t*
    energy team profile worksheet, 112*f*
    environmental and energy policy, 129*f*, 140*t*
    funded projects, 61–63*t*
    internal audit/self-inspection checklist, 105–109*f*
    legal compliance evaluation, 97*t*, 98–99*f*
    management review, 117*f*, 120–123*f*, 141–142*f*, 157–160
    monitoring at, 87*f*, 90*f*
    nonconformities, corrective/preventive action process, 113*f*
    objectives and target action plan, 59*f*, 137*f*
    operational controls, 77–78*t*, 79*f*
    procurement plan, 84*f*
    roles and responsibilities, 130–131*f*
    sample payback estimate, 61*f*
    significant energy users (SEUs), 36*t*
    team meeting agenda, 14*f*
    team meeting minutes, 15–16*f*
    utility costs, 32*f*

R

regular/preventive maintenance and cleaning, energy efficiency and, 42
requirements, EnMS, 7–20
    general, 7–8
    management representative, 10–13
    management responsibilities, 8–10
    team meetings, best practice uses for, 13–16, 14*f*, 15–16*f*
    team meetings, summary of, 16, 17–20*t*

S

significant energy users (SEUs), 4, 28–29, 30*f*
    identifying energy variables for, 43, 44–46*t*
    stratifying, 35*t*
SMART (specific, measurable, actionable, reviewable and relevant, and time based), 51, 53
space heaters, energy efficiency and, 41–42
specific, measurable, actionable, reviewable and relevant, and time based (SMART), 51, 53
strengths, weaknesses, opportunities, and threats (SWOT) analysis, 52
study/research type of objective, 52
sustaining EnMS stage, 5, 5*f*

T

thermostat settings, energy efficiency and, 40
training plan, 66–67, 68–69*t*

U

US Green Building Council, 79–80, 81

V

vending machine controls, energy efficiency and, 42